Preserving Scientific Data On Our Physical Universe

A New Strategy for Archiving the Nation's Scientific Information Resources

Steering Committee for the Study on the Long-term Retention
of Selected Scientific and Technical Records of the Federal Government

Commission on Physical Sciences, Mathematics, and Applications

National Research Council

NATIONAL ACADEMY PRESS
Washington, D.C.   1995

NOTICE: The project that is the subject of this report was approved by the Governing Board of the National Research Council, whose members are drawn from the councils of the National Academy of Sciences, the National Academy of Engineering, and the Institute of Medicine. The members of the committee responsible for the report were chosen for their special competences and with regard for appropriate balance.

This report has been reviewed by a group other than the authors according to procedures approved by a Report Review Committee consisting of members of the National Academy of Sciences, the National Academy of Engineering, and the Institute of Medicine.

The National Academy of Sciences is a private, nonprofit, self-perpetuating society of distinguished scholars engaged in scientific and engineering research, dedicated to the furtherance of science and technology and to their use for the general welfare. Upon the authority of the charter granted to it by the Congress in 1863, the Academy has a mandate that requires it to advise the federal government on scientific and technical matters. Dr. Bruce Alberts is president of the National Academy of Sciences.

The National Academy of Engineering was established in 1964, under the charter of the National Academy of Sciences, as a parallel organization of outstanding engineers. It is autonomous in its administration and in the selection of its members, sharing with the National Academy of Sciences the responsibility for advising the federal government. The National Academy of Engineering also sponsors engineering programs aimed at meeting national needs, encourages education and research, and recognizes the superior achievements of engineers. Dr. Robert M. White is president of the National Academy of Engineering.

The Institute of Medicine was established in 1970 by the National Academy of Sciences to secure the services of eminent members of appropriate professions in the examination of policy matters pertaining to the health of the public. The Institute acts under the responsibility given to the National Academy of Sciences by its congressional charter to be an adviser to the federal government and, upon its own initiative, to identify issues of medical care, research, and education. Dr. Kenneth I. Shine is president of the Institute of Medicine.

The National Research Council was established by the National Academy of Sciences in 1916 to associate the broad community of science and technology with the Academy's purposes of furthering knowledge and advising the federal government. Functioning in accordance with general policies determined by the Academy, the Council has become the principal operating agency of both the National Academy of Sciences and the National Academy of Engineering in providing services to the government, the public, and the scientific and engineering communities. The Council is administered jointly by both Academies and the Institute of Medicine. Dr. Bruce Alberts and Dr. Robert M. White are chairman and vice chairman, respectively, of the National Research Council.

Support for this project was provided by the National Archives and Records Administration (under Contract No. NAMA-S-92-0019), the National Oceanic and Atmospheric Administration (under Contract No. 50-DGNE-3-00105), and the National Aeronautics and Space Administration (under Contract No. S-54040-Z). The views expressed in this report are those of the authors and do not necessarily reflect the views of the sponsoring agencies or subagencies.

Library of Congress Catalog Card Number 94-68991
International Standard Book Number 0-309-05186-X

Additional copies of this report are available from:

National Academy Press
2101 Constitution Ave., NW
Box 285
Washington, DC 20055
800-624-6242
202-334-3313 (in the Washington Metropolitan Area)

B-499

Copyright 1995 by the National Academy of Sciences. All rights reserved.

Printed in the United States of America

**STEERING COMMITTEE FOR THE STUDY ON THE
LONG-TERM RETENTION OF SELECTED SCIENTIFIC AND TECHNICAL RECORDS
OF THE FEDERAL GOVERNMENT**

JEFF DOZIER, University of California, Santa Barbara, *Chair*
SHELTON ALEXANDER, Pennsylvania State University
MARJORIE COURAIN, Consultant (*deceased,* January 14, 1994)
JOHN A. DUTTON, Pennsylvania State University
WILLIAM EMERY, University of Colorado
BRUCE GRITTON, Monterey Bay Aquarium Research Institute
ROY JENNE, National Center for Atmospheric Research
WILLIAM KURTH, University of Iowa
DAVID LIDE, Consultant, Gaithersburg, Maryland
B.K. RICHARD, TRW
JOAN WARNOW-BLEWETT, American Institute of Physics

National Research Council Staff

Paul F. Uhlir, Associate Executive Director, Commission on Physical Sciences, Mathematics, and Applications
Mark David Handel, Program Officer, Board on Atmospheric Sciences and Climate
Alice Killian, Research Associate, Commission on Geosciences, Environment, and Resources
James E. Mallory, Staff Officer, Computer Science and Telecommunications Board
Scott T. Weidman, Senior Program Officer, Board on Chemical Sciences and Technology
Julie M. Esanu, Research Assistant, Commission on Physical Sciences, Mathematics, and Applications
David J. Baskin, Project Assistant, Commission on Physical Sciences, Mathematics, and Applications

COMMISSION ON PHYSICAL SCIENCES, MATHEMATICS, AND APPLICATIONS

RICHARD N. ZARE, Stanford University, *Chair*
RICHARD S. NICHOLSON, American Association for the Advancement of Science, *Vice Chair*
STEPHEN L. ADLER, Institute for Advanced Study
SYLVIA T. CEYER, Massachusetts Institute of Technology
SUSAN L. GRAHAM, University of California at Berkeley
ROBERT J. HERMANN, United Technologies Corporation
RHONDA J. HUGHES, Bryn Mawr College
SHIRLEY A. JACKSON, Department of Physics
KENNETH I. KELLERMANN, National Radio Astronomy Observatory
HANS MARK, University of Texas at Austin
THOMAS A. PRINCE, California Institute of Technology
JEROME SACKS, National Institute of Statistical Sciences
L.E. SCRIVEN, University of Minnesota
A. RICHARD SEEBASS III, University of Colorado
LEON T. SILVER, California Institute of Technology
CHARLES P. SLICHTER, University of Illinois at Urbana-Champaign
ALVIN W. TRIVELPIECE, Oak Ridge National Laboratory
SHMUEL WINOGRAD, IBM T.J. Watson Research Center
CHARLES A. ZRAKET, MITRE Corporation (retired)

NORMAN METZGER, Executive Director
PAUL F. UHLIR, Associate Executive Director

Preface

In January 1992 the National Archives and Records Administration (NARA) sponsored a three-day planning meeting at the National Research Council (NRC) to review the issues related to the long-term retention of the federal government's scientific and technical data in the physical sciences. The planning meeting was organized by the NRC's Commission on Physical Sciences, Mathematics, and Applications and provided the basis for this study, which was initiated in the fall of 1992 at the request of NARA. The National Oceanic and Atmospheric Administration (NOAA) and the National Aeronautics and Space Administration (NASA) subsequently provided additional support.

The study's steering committee, in consultation with the sponsors, developed the following charge to guide the writing of this report:

- Describe the status and plans for the government's archiving of observational and experimental data in the physical sciences. Identify the principal scientific, technical, information management, and institutional issues regarding the permanent archiving of such data.
- Assess the commonalities and differences among the case studies provided by the panels organized under this study (see below) in order to determine the extent to which common long-term retention policies and appraisal guidelines can be applied to disciplines that collect observational and experimental data in the physical sciences.
- Establish a set of goals, principles, and priorities, as well as generic retention criteria and appraisal guidelines that NARA can incorporate into its mission, program, and budget planning.
- Suggest mechanisms and processes for NARA and NOAA to use in implementing a program of data appraisal, retention, and preservation, and later in evaluating the effectiveness of the program.
- Provide a summary of findings, conclusions, and recommendations.

The steering committee formed five panels—in space sciences, atmospheric sciences, ocean sciences, geosciences, and physics, chemistry, and materials sciences—to provide their views on the key data retention issues from different disciplinary perspectives in the physical sciences. These panels each met twice and produced a set of working papers, which are published separately in *Study on the Long-term Retention of Selected Scientific and Technical Records of the Federal Government: Working Papers* (National Academy Press, Washington, D.C., 1995). The work of the panels was invaluable to the steering committee in framing the issues, in forming its conclusions and recommendations, and in producing its final report.

There are several aspects regarding the scope and focus of this report that should be mentioned. The committee devoted most of its attention to data stored on electronic media, rather than on paper or on other media. Almost all data are now acquired, stored, and distributed electronically. Thus, the preponderance of data archiving problems and their solutions must be considered in this context. Nevertheless, much of the advice offered here is equally relevant to data in other formats.

The principal focus of this report is on the long-term retention of data in the physical sciences. Much of the discussion, however, includes near-term data management issues, because effective archiving begins when the plans for acquiring a data set are made and extends throughout the life cycle of the data. Although the focus is exclusively on data in the physical sciences, the committee believes that the distinctions it has drawn between the experimental and the observational data, as well as the data management principles it has provided, are broadly applicable to most data in the other natural sciences. In addition, the strategic approach adopted by the committee necessarily involves all federal agencies that acquire and manage physical science data, and not simply the three agencies that sponsored this study.

Finally, it is necessary to point out that the committee was unable to achieve consensus on one major recommendation of the study, namely, the proposal to establish the National Scientific Information Resource (NSIR) Federation. Appendix B contains the minority opinion of the dissenting committee member, Roy Jenne. The rest of the committee members, who strongly support the NSIR Federation recommendation, are disappointed by this lack of unanimity and consider many of the assertions in the minority opinion to be based on an erroneous interpretation of what the report actually states or recommends. We leave that to the reader to judge. Nevertheless, we believe that the minority opinion can perhaps serve a useful purpose by drawing greater attention to these issues and by broadening the discussion of them among the sponsors of the study, the other science agencies, and the research community.

In conclusion, the committee hopes that its advice will help bring about the changes necessary to effectively preserve the valuable scientific data on our physical universe.

Jeff Dozier
Steering Committee Chair

Paul F. Uhlir
Study Director

Acknowledgments

The steering committee is very grateful to the many individuals who played a significant role in the completion of this study, including the members of the five ad hoc panels that provided conclusions and recommendations on data archiving from the different physical science disciplines; the individuals who briefed the steering committee and panels; and members of the National Research Council (NRC) staff who worked on various aspects of this study. The steering committee also extends its thanks to Trudy Peterson and Kenneth Thibodeau of the National Archives and Records Administration (NARA), William Turnbull and Helen Wood of the National Oceanic and Atmospheric Administration (NOAA), and Joseph King of the National Aeronautics and Space Administration (NASA), from the study's sponsoring agencies.

Gerd Rosenblatt, of Lawrence Berkeley Laboratory, chaired the Physics, Chemistry, and Materials Sciences Data Panel. The members were R. Stephen Berry, University of Chicago; Edward Galvin, The Aerospace Corporation; J.G. Kaufman, The Aluminum Association; Kirby Kemper, Florida State University; David R. Lide, Jr., consultant; and Edgar Westrum, Jr., University of Michigan. The steering committee gratefully acknowledges the detailed briefings and information provided to this panel by Donald Alderson, Department of Defense Nuclear Information Analysis Center; Frank Biggs, Sandia National Laboratories; Robert Billingsley, Defense Technical Information Center; Mark Conrad, NARA; Suzanne Leech, Bionetics, Inc.; Victoria McLane, Brookhaven National Laboratory; and Patricia Schuette, Battelle Pacific Northwest Laboratory.

The Space Sciences Data Panel was chaired by Christopher Russell of the University of California at Los Angeles. The panel members were Guiseppina Fabbiano, Harvard-Smithsonian Center for Astrophysics; Sarah Kadec, consultant; William Kurth, University of Iowa; Steven Lee, University of Colorado; and R. Stephen Saunders, Jet Propulsion Laboratory. The steering committee extends its thanks for the assistance of the following individuals, who provided briefings and other information to the Space Sciences Data Panel: Joe Allen, National Geophysical Data Center; Steven Blair, Los Alamos National Laboratory; Joseph Bredekamp, NASA; Dean Bundy, Naval Research Laboratory; David deYoung, National Optical Astronomy Observatories; Robert Frederick, Air Force Space Forecast Center; Joseph King, National Space Science Data Center; Knox Long, Space Science Telescope Institute; Guenther Riegler, NASA Astrophysics Division; Thomas Smith and Jud Stailey, Air Force Environmental Technical Applications Center; Earl Tech, Los Alamos National Laboratory; Raymond Walker, University of California at Los Angeles; and James Willet, NASA Space Physics Division.

Werner Baum, of Florida State University, was the chair of the Atmospheric Sciences Data Panel. The members were Marjorie Courain, consultant (deceased, January 14, 1994); William Haggard, Climatological Consulting Corporation; Roy Jenne, National Center for Atmospheric Research; Kelly Redmond, Desert Research Institute; and Thomas Vonder Haar, Colorado State University. The steering committee gratefully acknowledges the diverse and substantial inputs provided by the following individuals to the Atmospheric Sciences Data Panel: Larry Baume, NARA; Thomas Boden, Carbon Dioxide Information and Analysis Center; Dean Bundy, Naval Research Laboratory; Donald Collins, NASA; Richard Davis, National Climatic Data Center, P.C. Hariharan, Johns Hopkins University; and Gerald Stokes, Pacific Northwest Laboratories.

The Ocean Sciences Data Panel was chaired by Bruce Gritton, Monterey Bay Aquarium Research Institute. The members were Richard Dugdale, University of Southern California; Thomas Duncan, University of California at Berkeley; Robert Evans, Rosenstiel School of Marine and Atmospheric Science; Terrence Joyce, Woods Hole Oceanographic Institution; and Victor Zlotnicki, Jet Propulsion Laboratory. The steering committee extends its thanks for the briefings and other information provided to the Ocean Sciences Data Panel by Larry Baume, NARA; Donald Collins and Susan Digby, Jet Propulsion Laboratory; Ronald Fauquet, NOAA; Ted Tsui, Naval Research Laboratory; and R.S. Winokur, Office of Naval Research.

The Geoscience Data Panel was chaired by Theodore Albert, a private consultant. The members were Shelton Alexander, Pennsylvania State University; Sara Graves, University of Alabama in Huntsville; David Landgrebe, Purdue University; and Soroosh Sorooshian, University of Arizona. The steering committee gratefully acknowledges the information provided at the meetings of the Geosciences Data Panel by the following individuals: Roger Barry, National Snow and Ice Data Center; Daniel Cavanaugh, U.S. Geological Survey; Donald Collins, Jet Propulsion Laboratory; Katrin Douglass, Southern California Earthquake Center Data Center; William Draegar, U.S. Geological Survey; John Dwyer, NARA; Claire Henson, National Snow and Ice Data Center; Herb Meyers, National Geophysical Data Center; Ron Weaver, National Snow and Ice Data Center; and Thomas Yorke, U.S. Geological Survey.

Finally, the steering committee is grateful to the staff of the National Research Council: Paul F. Uhlir, associate executive director of the Commission on Physical Sciences, Mathematics, and Applications, who served as study director; Mark David Handel and Theresa Fisher (Board on Atmospheric Sciences and Climate), Alice Killian (Commission on Geosciences, Environment, and Resources), James E. Mallory (Computer Science and Telecommunications Board), and Scott T. Weidman and Taña Spencer (Board on Chemical Sciences and Technology), who provided staff support for the five panels; Julie M. Esanu, for the program assistance provided to the steering committee and panels and for the preparation of the final manuscript; David Baskin, for his work on preparing the final manuscript; Liz Panos, for coordinating the report review; and Roseanne Price, who edited the final manuscript.

Contents

SUMMARY	1
1 INTRODUCTION	10
Imperatives for Preserving Data on Our Physical Universe, 11	
A New Future for Scientific Data, 12	
2 THE CHALLENGE: PRESERVATION AND USE OF SCIENTIFIC DATA	13
Experimental Laboratory Data, 13	
Observational Data in the Physical Sciences, 15	
Summary of Major Issues, 29	
3 RETENTION CRITERIA AND THE APPRAISAL PROCESS	33
Retention Criteria, 33	
Other Elements of the Appraisal Process, 39	
Recommendations, 40	
4 THE OPPORTUNITIES: THE RELATIONSHIP OF TECHNOLOGICAL ADVANCES TO NEW DATA USE AND RETENTION STRATEGIES	42
Enabling Technologies and Related Developments, 43	
Opportunities for New Organizational Structures, 47	
5 A NEW STRATEGY FOR ARCHIVING THE NATION'S SCIENTIFIC AND TECHNICAL DATA	49
Fundamental Principles for Long-term Data Retention, 50	
The Proposed National Scientific Information Resource Federation, 51	
Recommendations for the Creation of the NSIR Federation, 55	
Recommendations Specifically for NARA, 57	
Recommendations Specifically for NOAA, 59	
REFERENCES	62
APPENDIX A List of Acronyms	64
APPENDIX B Minority Opinion	66

This study is dedicated in fond memory of Marjorie Courain.

Summary

Scientific data reflect both the organization and the chaos of the natural world. They stimulate us to develop concepts, theories, and models to make sense of the patterns they represent. The resulting abstractions are the formal and systematic ideas that constitute the understanding of relationships between causes and consequences, and perhaps may enable prediction of future sequences of events. Because scientists transform data from the material world into ideas, the observations of objects and processes in the physical world are the stimuli of scientific thought. Data are thus the seeds of scientific ideas.

There are strong motivations for preserving scientific observations:

- Many observations about the natural world are a record of events that will never be repeated exactly. Examples include observations of an atmospheric storm, a deep ocean current, a volcanic eruption, and the energy emitted by a supernova. Once lost, such records can never be replaced.
- Observed data provide a baseline for determining rates of change and for computing the frequency of occurrence of unusual events. They specify the observed envelope of variability. The longer the record, the greater our confidence in the conclusions we draw from it.
- A data record may have more than one life. As scientific ideas advance, new concepts may emerge—in the same or entirely different disciplines—from study of observations that led earlier to different kinds of insights. New computing technologies for storing and analyzing data enhance the possibilities for finding or verifying new perspectives through reanalysis of existing data records. Thus, the relative importance of data, both current and historical, can change dramatically, often in entirely unanticipated directions.
- The substantial investments made to acquire data records justify their preservation. The cost of preservation will almost always be small in comparison with the cost of observation. Because we cannot predict which data will yield the most scientific benefit in years ahead, the data we discard today may be the data that would have been invaluable tomorrow.

The assembled record of observational data thus has dual value: it is simultaneously a history of events in the natural world and a record of human accomplishment. The history of the physical world is an essential part of our accumulating knowledge, and the underlying data form a significant part of that heritage. They also portray a history of our scientific and technological development.

There are numerous socioeconomic reasons, in addition to the compelling scientific and historical motivations, for the long-term retention of observational, as well as certain types of experimental, data. For example, historical climate data have had well-documented uses in a broad range of applications in the manufacturing, energy, agriculture, transportation, communications, engineering, construction, insurance, and entertainment sectors. Such applications are common as well for other types of observational data on the Earth's environment. Experimental data in the physical sciences also have many industrial and other practical uses.

Today we can foresee the possibility of using the national resource of scientific data more advantageously than ever before as technological advances open new vistas for managing scientific information. Advances in data storage technologies make the long-term retention of virtually all data both feasible and affordable. The existence of the Internet and of the emerging National Information Infrastructure (NII) enables nationwide sharing and application of data that reside in appropriately configured databases.

Our new power to store, distribute, and access data and information is changing the way we work and think. However, the communities involved in the creation, retention, and use of scientific data about the physical world are not optimally organized. They commonly work toward disparate goals, are not well connected, and do not take full advantage of technological and conceptual advances in data management and communication. An entirely new approach to the long-term preservation of scientific data is now both feasible and essential. It must take advantage of advancing technology and of distributed communications and management structures to empower both the creators and the users of such data.

This study, performed at the request of the National Archives and Records Administration (NARA), and partially supported by the National Oceanic and Atmospheric Administration (NOAA) and the National Aeronautics and Space Administration (NASA), identifies the major issues regarding efforts to archive and use data in the physical sciences, establishes retention criteria and appraisal guidelines for those data, reviews important technological advances and related opportunities, and proposes a new strategy to help ensure access to the data by future generations.

THE CHALLENGE OF EFFECTIVE PRESERVATION AND USE OF SCIENTIFIC DATA

The results of scientific research are disseminated in this country through a hybrid system that includes professional society and other not-for-profit publishers, the commercial sector, and the government. The formal journals are published largely by the professional society and commercial sectors, while government agencies manage less formal reports (gray literature). Secondary abstracting and indexing services provide access to this literature, increasingly by electronic means. While there are strains in this system because of rising costs, increasing workload, and issues related to the protection of intellectual property, it has served U.S. science well and has been an invaluable link in the process of translating scientific advances into further advances, useful technology, and economic benefits.

The current system, however, is not well suited to handle the scientific and technical electronic databases that are the focus of this study. The cost of maintaining these databases is typically too great to be covered by user fees; instead these databases must be considered part of the national scientific heritage. Some government agencies have accepted responsibility for maintaining and disseminating the data resulting from their research and development. In some cases, this system is working reasonably well, but in others there are problems even with providing current access. Archiving for the long term raises questions in all cases, however.

A general problem prevalent among all scientific disciplines is the low priority attached to data management and preservation by most agencies. Experience indicates that new research projects tend to get much more attention than the handling of data from old ones, even though the payoff from optimal utilization of existing data may be greater.

With regard to laboratory data, government programs have existed since the 1960s to compile results from the world scientific literature, to check the data carefully, and to prepare databases of critically evaluated data. Despite chronic underfunding, these programs have produced databases of lasting value to the nation, and the government investment in creating and maintaining these databases has been repaid many times over.

In the area of observational databases, the situation is mixed. Federal agencies collect large amounts of observational data, which in many cases are continuously added to the available record of Earth and space processes. The data sets resulting from these activities are sometimes well-documented and maintained in readily accessible form; in many other cases, however, while the data are saved, they are exceedingly difficult or impossible to access or use, and thus are effectively unavailable.

The most important deficiencies are in the documentation, access, and long-term preservation of data in usable form. Insufficient documentation is a generic problem that affects, in varying degrees, all the classes of data addressed in this study. Furthermore, few of the federal data centers can give adequate attention to long-term archiving because they are stretched thin by current demands and inadequate resources. Even the data that are archived may become inaccessible because they are not regularly migrated to new storage media as the hardware and software used to access the data become obsolete or inoperable.

Another major problem inhibiting access to data is the lack of directories that describe what data sets exist, where they are located, and how users can access them. In many cases the existence of the data is unknown outside the original scientific groups, and even if known, there frequently is not enough information for a potential user to assess their relevance and usefulness. The lack of adequate directories adversely affects the exploitation of our national data resources and leads to unnecessary duplication of effort.

A significant fraction of the archived scientific data is held by the federal agencies that collected the data as part of their mission. However, a large amount of valuable scientific data gathered with federal funds is never archived or made accessible to anyone other than the original investigators, many of whom are not government employees. In many instances, the organizations and individuals that receive government contracts or grants for scientific investigations are under no obligation to retain the data collected, or to place them in an accessible archive at the conclusion of the project. Thus, data sets that commonly are gathered at great expense and effort are not broadly available and ultimately may be lost, squandering valuable scientific resources and much of the public investment spent in acquiring them. Clearly, there is a great need for the agencies to get more return on their investment in science by the simple expedient of making the data collected under their auspices accessible to others.

Finally, the holdings of scientific and technical data by NARA in electronic or any other form are very small in comparison with the data holdings of the federal agencies and the organizations supported by them. Moreover, NARA's budget for its Center for Electronic Records, which has the formal responsibility for archiving all types of federal electronic records, was only $2.5 million in FY 1994, a budget lower than that of many of the individual agency data centers reviewed by the committee in this study. Given NARA's current and projected level of effort for archiving electronic scientific data, it is obvious that NARA will be unable to take custody of the vast majority of these scientific data sets. Therefore, a coordinated effort involving NARA, other federal agencies, certain nonfederal entities, and the scientific community is needed to preserve the most valuable data and ensure that they will remain available in usable form indefinitely. The challenge is to develop data management and archiving procedures that can handle the rapid increases in the volumes of scientific data, and at the same time maintain older archived data in an easily accessible, usable form. An important part of this challenge is to persuade policymakers that scientific data and information are indeed a precious national resource that should be preserved and used broadly to advance science and to benefit society.

RETENTION CRITERIA AND THE APPRAISAL PROCESS

The National Archives and Records Administration appraises records on the basis of their informational and evidential value. It is concerned with records of long-term value, those records that will probably have value long after they cease to have immediate, or primary, uses. The value of scientific and technical data is primarily informational and is based on the scientific content of the records, rather than on the evidence they provide concerning the activities of the agency that collected or created them.

Recommendations

The recommendations below regarding the retention criteria and appraisal process should be applied—by those responsible for stewardship—to all physical science data. Similar criteria and appraisal guidelines must be developed for data in other disciplines. This is a topic of primary concern not only to NARA, NOAA, and NASA, but to all scientists, data managers, and archivists who work with such records.

As a general rule, all observational data that are nonredundant, useful, and documented well enough for most primary uses should be permanently maintained. Laboratory data sets are candidates for long-term preservation if there is no realistic chance of repeating the experiment, or if the cost and intellectual effort required to collect and validate the data were so great that long-term retention is clearly justified. For both observational and experimental data, the following retention criteria should be used to determine whether a data set should be saved: uniqueness, adequacy of documentation (metadata), availability of hardware to read the data records, cost of replacement, and evaluation by peer review. Complete metadata should define the content, format or representation, structure, and context of a data set.

The appraisal process must apply the established criteria while allowing for the evolution of criteria and priorities and must be able to respond to special events, such as when the survival of data sets is threatened. All stakeholders—scientists, research managers, information management professionals, archivists, and major user groups—should be represented in the broad overarching decisions regarding each class of data. The appraisal of individual data sets, however, should be performed by those most knowledgeable about the particular data—primarily the principal investigators and project managers. In some cases, they may need to involve an archivist or information resources professional to assist with issues of long-term retention.

Classified data must be evaluated according to the same retention criteria as unclassified data in anticipation of their long-term value when eventually declassified. Evaluation of the utility of classified data for unclassified uses needs to be done by stakeholders with the requisite clearances to access such data.

OPPORTUNITIES CREATED BY TECHNOLOGICAL ADVANCES FOR NEW DATA USE AND RETENTION STRATEGIES

Rapid progress in information technology continually alters both the quantity and the quality of scientific information and periodically stimulates fundamental modification of data management and archiving strategies. Recent technological advances have enabled new methods and strategies for data storage and retrieval and have created better ways of connecting users to data resources and to each other. Moreover, the evolving technologies are catalysts for revising organizational structures to manage distributed scientific data archives much more effectively.

Table S.1 provides a summary of new technologies and related developments that enable a new strategy for the management of scientific and technical data. These advances in information technologies

TABLE S.1 New Technologies and Related Developments That Enable a New Strategy for the Management of Scientific and Technical Data

New Technology Trends and Related Developments	Key Features	What Is Enabled?
High-performance computer networks	Distributed functions; rapid delivery of large data volumes	Location of databases and archives where best managed; collaborative work; distributed organizations; distributed responsibility
Low and declining cost of storage	Inexpensive backup; continually declining cost; ease of migration	Deferral of archiving decisions; trust in distributed management due to safe storage backup
Advanced data management	Ability to rigorously and formally manage diverse data types	More complex data structures (other than "flat files") handled in archives, with great potential advantages
Changing requirements for information technology professionals	Ability of personnel with lower technical skills to succeed in data management roles	Ability to entrust scientific data management in a distributed environment
High reliability of technology components	Availability of better components and connections; reduced procurement and operations costs	Reduced cost and effort in data migration; trusted connections for communication and collaboration
Development and acceptance of standards	Agreement on terms, interfaces, media, procedures	Reduced effort to communicate and apply results of others; ability to concentrate on mission issues and not on technology support

and data management support the creation of a highly distributed, federated management structure for our nation's scientific information resources.

A NEW STRATEGY FOR ARCHIVING THE NATION'S SCIENTIFIC AND TECHNICAL DATA

In order to respond adequately to the imperatives for preserving data about the physical universe and to take advantage of the technological advances described above, the federal government should create an integrated and adaptive infrastructure and related processes for providing ready access to the national resource of scientific and technical data and related information. Such an effort must support the needs of data originators, users, and custodians across all phases of the data life cycle, from origin to use by future generations. The committee believes that the following principles should guide the effort of the government agencies in the long-term retention of scientific and technical data:

- *Data are the lifeblood of science and the key to understanding this and other worlds. As such, data acquired in federal or federally funded endeavors, which meet established retention criteria, are a critical national resource and must be protected, preserved, and made accessible to all people for all time.*
- *The value of scientific data lies in their use. Meaningful access to data, therefore, merits as much attention as acquisition and preservation.*

- *Adequate explanatory documentation, or metadata, can eliminate one of today's greatest barriers to use of scientific data.*
- *A successful archive is affordable, durable, extensible, evolvable, and readily accessible.*
- *The only effective and affordable archiving strategy is based on distributed archives managed by those most knowledgeable about the data.*
- *Planning activities at the point of data origin must include long-term data management and archiving.*

The Proposed National Scientific Information Resource Federation

The committee believes that the federal government should create a National Scientific Information Resource Federation—an evolutionary and collaborative network of scientific and technical data centers and archives—to take on the challenge of providing effective access to and preservation of important data and related information. Such an initiative would begin to exploit fully our nation's significant investment in the physical (and other) sciences and the data acquired with that investment. Several critical concepts must govern any federated management structure for it to function properly (Handy, 1992):

- **Subsidiarity**—the power is assumed to lie with the subordinate units of an organization. Power can be relinquished, but not taken away. The subordinate units typically are best qualified to make operational decisions that directly affect them and that they will be implementing. The central management is allowed only those powers needed to ensure that the subordinates do not damage the organization. It is clear that the strengths of the current system for managing scientific and technical data and information in the United States are distributed among a number of diverse data centers and archives, both within and outside the government. A successful federation of these existing institutions would recognize that they are the locations of expertise on their respective data holdings. Thus the central organization should be small and should not micromanage the day-to-day operations of the subsidiary organizations.
- **Pluralism**—the members are interdependent. In a federation, the individual subsidiary organizations recognize the advantages of belonging to the federation, because of products or services that can be obtained from other elements in the federation. The existence of many specialized data centers and archives, as well as the possibility of creating new ones in a networked environment, can offer significant economies of scale and improved sharing of ideas and expertise. What is good for the subsidiary element also should be good for the whole. Pluralism, coupled with subsidiarity, guarantees a measure of democracy in the federation.
- **Standardization**—interdependence requires compatible languages, communications, basic rules of conduct, and units of measurement. These elements may be summarized as technical and procedural standardization. Standards that are developed by consensus of the subsidiary elements (e.g., the participating data centers, archives, and researchers) are widely recognized as essential to the successful management of data.
- **Separation of powers** (responsibilities)—a system of checks and balances is necessary to ensure that the central authority does not take on unnecessary power. This principle must be incorporated into the federation's organizational structure.
- **Strong leadership**—the central coordinating element or executive office must act as the standard bearer, promoting the federation's established goals and objectives while reminding the subsidiary organizations of the importance of carrying out their responsibilities.

A federated data management system would be consistent with the goal of the National Information Infrastructure to distribute information resources broadly throughout our society. The technology is available to make a fully networked, but highly distributed system of data centers and archives both feasible and desirable. Such a system would be efficient in providing access to scientific data and information to a large number of potential users and would maximize the government's return on the very large investment that initially went into acquiring those data. From an organizational standpoint, a federated management structure would allow the disparate elements to continue to specialize in what they each do best and to fulfill their individual organizational mandates, while providing some efficiencies of scale and political leverage in addressing the most pressing issues. The committee believes this approach is especially timely and important in an era of federal government budget reductions.

Recommendations

The committee thus recommends that the federal government take the following steps for adequately preserving and providing access to data about our physical universe:

Adopt the National Scientific Information Resource (NSIR) Federation concept as an integral part of the National Information Infrastructure (NII). This concept must encompass not only an electronic network, but also individuals, organizations, communities, data resources, procedures, guidelines, and associated activities of data generation, management, custodianship, and use. The NSIR Federation thus should provide the means for defining a coherent approach to managing the life cycle of scientific data. This approach should be developed and implemented through consensus of collaborating organizations with diverse and autonomous missions. The interagency Global Change Data and Information System is an example of a prototype NSIR Federation, focused on data for a specific set of interdisciplinary science problems. The NSIR Federation would build on such efforts, providing for better coordination and interaction among them, and would help organize fledgling efforts to preserve and provide broad access to data in other disciplines.

The administration should take the steps necessary to fully define and create the NSIR Federation. There are at least two potential focal points within the administration for planning such an activity. These are the interagency Information Infrastructure Task Force for the NII and the National Science and Technology Council. A convocation of representatives from the scientific, data and information management, and archiving communities would be a good way to help define and inaugurate this initiative.

Following the formal authorization by the federal government for creating the NSIR Federation, the principal parties, including NARA and NOAA, should conclude agreements for the implementation of a distributed archive system. The system should involve all relevant institutions, including nongovernmental entities that are funded by the federal government or that maintain data that were acquired with federal funds. As a general principle, data collected by an agency should remain with that agency indefinitely. The committee recognizes that this recommendation may require significant operational changes for agencies other than NOAA, and even some changes with respect to NOAA's data activities. Furthermore, the associated agencies in the NSIR Federation must work together, under the lead of a small executive office with the expertise to establish data management guidelines and minimum criteria for adequate metadata that could be applied across the entire Federation. The executive office could be either a high-level interagency coordinating committee or a new office at an appropriate federal agency, such as the National Science Foundation, which has a broad scientific and technical as well as communication mandate. In any case, the executive office should resist the typical tendency toward bureaucratic accretion of power, personnel, and resources, as well as the tendency to consolidate and centralize data holdings. A management council consisting of representatives of the member organizations should be created to help ensure that the executive office function remains fully responsive to all members of the federation.

Data access and preservation services should be implemented on the most cost-effective basis possible for the Federation. For example, one institution should provide a service to one or more other institutions in order to exploit potential economies of scale and focal points of expertise. This measure might increase the cost to the providing institution, but would decrease the overall cost to the federation, the government, and the taxpayer.

The institutions belonging to the NSIR Federation should develop a process for collaborating effectively on specific initiatives. This process should provide a mechanism to define and prioritize data management and preservation initiatives, to establish the required agreements between collaborating organizations, and to secure funding for each initiative. Each participating organization would contribute to the federation according to its particular strengths and in a manner consistent with the founding charter. In addition, an independent advisory board consisting of experts from user groups should be formed in support of each initiative.

The NSIR Federation should develop a national resource of information technology that is consistent with its chartered objectives and that can be effectively distributed to institutions that must manage data. These technologies would include complete products, designs, guidelines, standards, and methodologies. A related long-term technology strategy, or "technology navigation" function, should be developed to help guide these efforts.

The NSIR Federation should institute an independently managed process for awarding NSIR certification to member scientific institutions and their data and information systems on the basis of well-defined criteria and standards. The certification process should be managed by a nongovernmental, not-for-profit organization, which would receive technical guidance from the participating federal agencies. The certification needs to have credibility in the community, so that nonmember institutions will aspire to attain certification and have it tagged to their products. The certification also should be something that commercial value-added providers seek to increase the credibility of their products.

It also is important for the committee to state what the NSIR Federation should not be. It should not become an expensive bureaucratic entity. The executive office must not impose any standards or information technologies from above that have not been validated through a consensus process of the member organizations. Finally, the executive office must not attempt to micromanage the operations of the participants, nor should it have any direct control over their budgets and funding allocations.

Recommendations Specifically for NARA

Although NARA has a legislative mandate to preserve federal records, it cannot today, nor will it likely ever be able to, act as the custodian of most physical science data. The data volume is too great in relation to the very low funding appropriated to NARA, the NARA staff do not have the specialized scientific knowledge, the interagency linkages are not in place, and a huge infrastructure similar to that which already exists at other agencies would need to be duplicated by NARA. In addition, the designation of a federal record is sometimes irrelevant to the archival process for scientific and technical data, and many data of long-term interest do not meet the existing definition of a federal record.[*] Hence,

---

[*] "'[Federal] records' includes all books, papers, maps, photographs, machine readable materials, or other documentary materials, regardless of physical form or characteristics, made or received by an agency of the United States Government under Federal law or in connection with the transaction of public business and preserved or appropriate for preservation by that agency or its legitimate successor as evidence of the organization, function, policies, decisions, procedures, operations, or other activities of the Government or because of the informational value of the data in them" (44 U.S.C. 3301).

NARA has a special role as a partner in the archiving process for scientific and technical data sets that is different from its traditional role as the nation's archives.

The committee makes the following specific recommendations to NARA in addition to those made elsewhere in this report:

NARA should strengthen its liaison with each federal agency that produces scientific and technical data to ensure that appropriate attention is devoted to their long-term retention in a distributed storage environment.

NARA should form standing advisory committees with managers of scientific data, historians, and scientific researchers to address the retention and appraisal of scientific and technical data collections and related issues.

NARA should collaborate with other agencies that maintain long-term custody of data to develop an effective access mechanism to these distributed archives. The initial step should focus on locator systems and evolve toward a transparent access system.

Finally, NARA should work with the scientific community and potential sources of scientific data to develop adaptable *performance* criteria for data formats and media, rather than mandating narrow and inflexible product standards.

Recommendations Specifically for NOAA

As the largest holder of earth sciences data in the United States, NOAA has a vast amount of scientific data stored at a number of facilities across the country. NOAA thus has an especially important role in the preservation of our nation's observational data on the physical environment. The committee makes the following specific recommendations to NOAA:

NOAA should place a higher priority on documenting and establishing directories of its data holdings.

NOAA, with the active cooperation of NARA, should lead efforts to better define technology-independent standards for archiving, storing, and transmitting the data within its purview.

Finally, NOAA, as well as every other federal science agency, should ensure that:

- **all its data are shared and readily available;**
- **it fulfills its responsibility for quality control, metadata structures, documentation, and creation of data products;**
- **it participates in electronic networks that enable access, sharing, and transfer of data; and**
- **it expressly incorporates the long-term view in planning and carrying out its data management responsibilities.**

The creation of the committee's proposed NSIR Federation would help provide a collaborative mechanism and more sustained peer pressure to meet these objectives, and thus enhance the value of scientific and technical data and information resources to the nation.

1
Introduction

Standing at the intersection of past and future, we humans are fascinated with the events of yesteryear and intrigued with what tomorrow will bring. Our prehistoric ancestors began the process of recording aspects of the environment that were important to them (Marshack, 1985; Boorstin, 1992). Today we are curious about many more worlds, ranging from those of atomic size to those of cosmic scale. With instruments on Earth and in space, we seek to capture views of reality that will help us understand nature and our relationship to it.

Scientific data reflect both the organization and the chaos of the natural world. They stimulate us to develop concepts, theories, and models to make sense of the patterns they represent. The resulting abstractions are the product of scientific endeavor, the goal being to develop the formal and systematic ideas that constitute the understanding of relationships between causes and consequences and perhaps may enable prediction of future sequences of events. Because scientists transform data from the material world into ideas, the observations of objects and processes in the physical world are the stimuli of scientific thought. Data are thus the seeds of scientific ideas.

Science generally works by proceeding from data to understanding through a process of organizing the data and analyzing their implications. The following definitions, adapted from *Setting Priorities for Space Research: Opportunities and Imperatives* (NRC, 1992a), indicate how the process works:

- *Data* are numerical quantities or other factual attributes derived from observation, experiment, or calculation.
- *Information* is a collection of data and associated explanations, interpretations, or other textual material concerning a particular object, event, or process.
- *Knowledge* is information organized, synthesized, or summarized to enhance comprehension, awareness, or understanding.
- *Understanding* is the possession of a clear and complete idea of the nature, significance, or explanation of something; it is the power to render experience intelligible by ordering particulars under broad concepts.

This process is cyclical. New data confirm or refute existing theories and stimulate new understanding, which generates new and deeper questions that often need entirely new sets of observations to begin the process of answering them. New understanding also leads to increased technological capability, and that in turn makes new observations possible and again allows us to contemplate more sophisticated questions.

Thus observations and scientific progress are intertwined; data from the physical world ensure that science is founded on reality as we try to answer the unending "how" and "why" questions that are part of being human. The answers become understanding that enables us to develop schemes for predicting or not being surprised by future events. And understanding, we hope, ultimately leads to wisdom about our interactions with the world around us.

IMPERATIVES FOR PRESERVING DATA ON OUR PHYSICAL UNIVERSE

The scientific reasons for preserving data derive from the fact that observations, knowledge, and understanding are cumulative. Thus we believe that the more complete the record, the more we can extract from it.

Many observations about the natural world are a record of events that will never be repeated exactly. Examples include observations of an atmospheric storm, a deep ocean current, a volcanic eruption, and the energy emitted by a supernova. Once lost, such records can never be replaced.

Observed data provide a baseline for determining rates of change and for computing the frequency of occurrence of unusual events. The longer the record, the greater our confidence in the conclusions we draw from it. Our traditional observational records have portrayed frozen instants of reality. If preserved, they will continue to provide insights, but if neglected, they will melt away.

A data record is also worth preserving because it may have more than one life. As scientific ideas advance, new concepts emerge—in the same or entirely different disciplines—from study of observations that led earlier to different kinds of insights. New computing technologies for storing and analyzing data enhance the possibilities for finding or verifying new perspectives through reanalysis of existing data records. Thus, the relative importance of data, both current and historical, can change dramatically, often in entirely unanticipated directions. This means that the reanalysis of data, even in the distant future, may bring new understanding, which will again increase the value of those data over that which we might have assigned to them at the time of their archiving. Finally, the substantial investments made to acquire data records usually justify their preservation. The cost of preservation will almost always be small in comparison with the cost of observation. Because we cannot predict which data will yield the most scientific benefit in years ahead, the data we discard today may be the data that would have been invaluable tomorrow.

The assembled record of observational data thus has dual value: it is simultaneously a history of events in the natural world and a record of human accomplishment. The history of the physical world is an essential part of our accumulating knowledge, and the underlying data form a significant part of that heritage. They also portray a history of our scientific and technological development.

With appropriate explanatory documentation, often referred to as metadata, the data demonstrate the increasing sophistication of our attempts to understand our natural surroundings and the technological capabilities we apply to the task. Preserved for study by future generations, the data will speak across the years about what we tried to do, where we succeeded, and where we failed. With increasing capabilities for analyzing and conceptualizing patterns in data, those who follow may find in our archived data important clues that we could not or did not see. At the same time, our descendants will be grateful that we preserved a sufficiently long history of their world that they can make important decisions about their own future.

There are numerous socioeconomic reasons, in addition to the compelling scientific and historical motivations, for the long-term retention of observational, as well as certain types of experimental, data. For example, historical climate data have had well-documented uses in a broad range of applications in manufacturing, energy, agriculture, transportation, communications, engineering, construction, insurance, and entertainment (OTA, 1994). Such applications are common for other types of observational data on the Earth's environment. Experimental data in the physical sciences also have many industrial and other practical uses. Additional examples of the long-term uses of the various physical science data are provided in the next chapter.

A NEW FUTURE FOR SCIENTIFIC DATA

The collections of scientific data acquired with government and private support are the foundation for our understanding of the physical world and for our capabilities to predict changes in that world. In the years ahead, the volumes of those collections of data will increase dramatically. They will stimulate advances in our scientific understanding and in our applications of that understanding to pursue important national goals. The scientific data in federal, state, and private databases thus constitute a critical national resource, one whose value increases as the data become more readily and broadly available.

Today, we can foresee the possibility of using the national resource of scientific data more advantageously than ever before, as technological advances open new vistas for managing and accessing scientific information. Growing computational power enables new approaches to the analysis, management, and application of data. Advances in data storage technologies make the long-term retention of virtually all data both feasible and affordable. The existence of the Internet and of the emerging National Information Infrastructure (NII) enable unprecedented nationwide sharing and application of data that reside in appropriately configured databases. Automatic search procedures, file transfer capabilities, and the accelerating use of the World Wide Web functions on the Internet illustrate the power of the contemporary technology. It is important to note that these enabling technologies have emerged in a short time span; equally rapid advances can be anticipated in the years ahead, which will further facilitate the search for and access to the nation's data resources.

Our new power to store and distribute data and information is changing the way we work and think. However, the communities involved in the creation, retention, and use of scientific data about the physical world are not optimally organized. They commonly work toward disparate goals, are not well connected, and do not take full advantage of technological and conceptual advances in data management and communication. An entirely new approach to the long-term preservation of scientific data is now both feasible and essential. It must take advantage of advancing technology and of distributed communications and management structures to empower both the creators and the users of such data.

This study identifies the major issues regarding existing efforts to archive and use data in the physical sciences, establishes retention criteria and appraisal guidelines for those data, reviews important technological advances and related opportunities, and proposes a new strategy to ensure access to the data by future generations.

2
The Challenge:
Preservation and Use of Scientific Data

We advance our understanding of the physical universe by building on current and past studies in individual disciplines, by collecting and analyzing new types of data, and by using past observations in entirely new ways not envisioned when the data were initially collected. The more complete the record of scientific data and information, the more new understanding and knowledge we can extract from it. Observations of natural phenomena typically represent a record of events that will never be repeated in a dynamic universe that continually changes in time and varies in space. New scientific advances have had significant, sometimes profound, societal and economic impacts and may be expected to be equally important in the future. Scientific data and information are at the heart of these advances and are essential for new discoveries. Therefore, they constitute a precious national resource.

The sections that follow describe briefly the two major types of data that are of critical importance in the physical sciences—experimental laboratory data in physics, chemistry, and materials sciences, and observational data in the earth and space sciences. In each of these broad areas the progress that has been made to date in terms of long-term preservation and accessibility is characterized, and the key issues identified. More comprehensive descriptions of the status of long-term data retention in the various physical science discipline areas are in the volume of working papers prepared as background for this report (NRC, 1995).

EXPERIMENTAL LABORATORY DATA

The experimental sciences have progressed over the centuries by building on the concepts, theories, and factual information resulting from each generation of scientific inquiry. The observations of Tycho Brahe were used by Kepler to develop his laws of planetary orbits, and Newton's formulation of mechanics drew upon the previous work of Galileo, Kepler, and others. A century of measurements on properties of the chemical elements provided the raw material needed for Mendeleev to construct his periodic table. The history of science is rich in examples where the introduction of new, often revolutionary, concepts rested on data that had been preserved from previous scientific investigations. Furthermore, the technology of tomorrow is often based on the laboratory data of today or yesterday.

The explosive growth of science in this century provides many other examples of the key role of data from previous experiments. When Townes and Schawlow published their landmark 1958 paper that demonstrated the theoretical possibility of building a laser, intensive efforts were started to find a real physical system that would meet the necessary requirements. Data on atomic spectra, some of them 60 to 70 years old, provided the key to creation of the first working gas laser. If it had been necessary to make new measurements on every conceivable system in order to select the most promising for trial, the invention of the laser—and all the new technology and economic benefits that it has brought—would have been delayed for many years.

The crash program to improve rocket propulsion systems following the launch of the first Soviet Sputnik provides another example. Data on the thermodynamic properties of a wide range of substances were essential to the efforts to optimize rocket engine performance. A concerted government program was started to build a database of thermodynamic properties for rocket engine design. Although some new laboratory measurements were required, many of the needed data were in the scientific literature, some published as early as 1880. The availability of these older data significantly aided the rocket engine program.

Data generated by scientists and engineers in the fields of physics, chemistry, and materials science have traditionally been published in research journals, which serve both a current dissemination and an archival function. This journal system has served science well for 300 years. Many scientific libraries throughout the country provide access to these journals. Because back volumes are kept in libraries in many different places, there is little danger of irreparable loss from a natural catastrophe. Many scientific societies also have depository systems that allow authors to submit voluminous data sets that cannot be published in the journals because of lack of space. The societies maintain these archives, generally on microfilm, and supply copies on request.

While the growing use of electronic recording and storage techniques is already affecting the traditional journal system, we can expect publishers to take advantage of the new technology to meet new needs. Scientific societies are beginning to implement electronic archives for preserving data that are too voluminous to publish in paper formats. For example, the American Chemical Society recently began to make data from papers in its leading journal (*Journal of the American Chemical Society*) available on the Internet. It is a natural step from the paper and microfilm archives that such societies now maintain to the electronic archives of the future. Clearly, these private sector archives must be an integral part of the overall concept of a "National Scientific Information Resource."

Electronically recorded data in the laboratory physical sciences are of two forms, original experimental measurements and evaluated compilations of published data. These are examined here in turn.

Original Experimental Measurements

Recent decades have seen significant changes in the form of "original data." A raw experimental result was, in the past, typically a measured value such as a voltage or distance. The investigator read these measurements from instruments, wrote them in a notebook, treated them arithmetically to obtain the desired scientific variable from the raw measurement, and interpreted them. The original measurements were eventually discarded in most cases. Today, many raw data are acquired and processed electronically as soon as they are entered into the computer, so that only the processed data exist long enough for anyone to look at. With rapid, automated data acquisition and manipulation, the option exists to keep electronic data and reanalyze them as required. However, automated data collection often results in large volumes of insignificant data, so that in many experiments the data stream is screened and most of the data are discarded in real time by a computer program or by the experimenter. For example, spectroscopists used to keep, at least temporarily, the photographic plates or recorder charts from which they had taken measurements. Now the spectral features may be analyzed electronically immediately upon measurement, and only the attributes of relevant features are recorded. The fraction of the raw data that is saved after initial processing may be small, sometimes less than one part in 10,000. *In virtually all cases, there is no justification for preserving the raw data, because the experiment can be repeated in those rare instances in which an unanticipated future interest appears.*

When considering laboratory data of this kind, it is usually best to recognize that no one knows as much about the original data as the original experimenter. If the experimenter does not find the raw data worth preserving (and worth documenting), then the data are probably not going to be of use to anyone else. Because the number of stages of processing (e.g., replication, averaging, coordinate transformations, applying corrections, and so on) differ for every type of measurement and undergo continual evolution as new techniques are introduced, it would be fruitless to try to formulate generic retention criteria for all types of laboratory data.

However, there are certain classes of laboratory data (where "laboratory" is used in a broad sense) that should be candidates for preservation if properly documented, because it would be impossible or impractical to reproduce the measurements. Some of the data taken in large plasma physics facilities fall in this category, because reproduction of the facilities would be extremely costly. A more striking example is the spectroscopic and other measurements from nuclear tests in the atmosphere, which it is hoped will never be reproduced. On a more mundane level, properties of engineering materials, measured as a part of large government research and development programs, provide many data of possible interest in the future. Such data are acquired as a small step in a larger program and usually are not published in the scientific literature or disseminated by the usual channels. They would be costly to reproduce because many of the materials were specially prepared with unique fabrication technology. Examples include polymer and sensor data from the Strategic Defense Initiative, engineering data from the National Aeronautics and Space Administration (NASA), and the superconducting materials measurements carried out to develop magnet fabrication techniques for the canceled Superconducting Super Collider. Even though this project will not be completed, the materials measurements should be saved, because they may well be applicable to future engineering projects.

Evaluated Compilations

Compilations resulting from the critical analysis of a large body of data from the scientific literature are a separate area for consideration. Well-known examples include thermodynamic property compilations such as the National Institute of Standards and Technology's Joint Army-Navy-Air Force (JANAF) tables and the thermophysical properties disseminated by the Department of Defense's Center for Information and Data Analysis and Synthesis at Purdue University (see the Physics, Chemistry, and Materials Sciences Data Panel report in the NRC (1995) report for a detailed discussion of these examples). The Department of Energy operates several data evaluation centers in nuclear physics and chemistry. In such centers, the data and backup documentation are not impossible to replace; they simply represent so much effort and exercise of specialized scientific judgment that it would be extremely costly to redo the work. The cost of not having the data available, although usually difficult to measure other than anecdotally, can be much higher than the cost of preserving them. In particular, if it becomes necessary in the future to expand or extend the compilation, the full documentation (e.g., data extracted from references, fitting programs, notes on the analysis techniques, and the like) will provide a valuable base for the new work. A major concern in considering these data collections is how the data and the underlying documentation can be preserved and made accessible if the centers producing them lose their funding or expert personnel. This concern increases as government agencies downsize their activities.

OBSERVATIONAL DATA IN THE PHYSICAL SCIENCES

Over the past two decades, the National Research Council and other groups have issued numerous reports that have addressed data management issues, including long-term retention requirements, for digital observational data in the earth and space sciences (NRC, 1982, 1984, 1986a,b, 1988a,b, 1990, 1992b, 1993; GAO, 1990a,b; Haas et al., 1985; NAPA, 1991). Most of these reports have focused quite narrowly on the data management or archiving problems of specific disciplines or agencies, and none has addressed comprehensively the issues associated with the long-term retention of observational and experimental data in the physical sciences.

Major Characteristics of Observational Data

Observational data sets, like laboratory data, include digital information (in both written and electronic form), graphical records, and verbal descriptions. The records exist as ink on paper, punched paper, film (including microforms), magnetic tape of many types (including videotape), magnetic disk, and digital optical media (including CD-ROM). Over the past three decades, however, the dominant form of data collection and storage has been electronic.

Observational data can be characterized by the collection and management practices applied throughout the life cycle of their existence. One might characterize two major practices driven by the funding models for conducting the underlying science. The "big science" funding model creates a funding umbrella for multiple individuals and institutions to conduct coordinated data acquisition, investigation, and publication. Often, these large programs adopt a standard approach for life-cycle data management. However, there is usually little standardization among the big science programs. Examples of such programs include the World Ocean Circulation Experiment, the World Climate Research Program, and NASA's Mission to Planet Earth (CENR, 1994). The other funding model, "small science," funds individuals or small groups of individuals to conduct independent data acquisition, analysis, and publication. Typically, these investigators plan, design, and implement their own data management strategy with little interaction with the rest of the scientific community. The data generated under both models have long-term value, both for science and for the broader interests of the nation.

Specific subdisciplines also impose different requirements on long-term data management. For instance, while there is general agreement within the physical oceanography community on the definition of standard observation variables and the processes of measuring those variables, the same cannot be said for biological oceanography. Because of differences in measuring techniques, lack of community agreement on naming standards, and the scientific process by which biology progresses, data management for biological data sets is inherently more complex than in physical oceanography. The data from these two subdisciplines will have to accommodate multiple naming schemes and alternate taxonomies. Therefore, data managers and archivists have to deal with differing approaches and vocabularies among disciplines, evolution of discipline research paradigms over time, and diverging concepts and methods within a discipline.

Scientific research leads to the creation of data that can be processed and interpreted at different levels of complexity. Typically, each level of processing adds value to the original (level-0) data by summarizing the original product, synthesizing a new product, or providing an interpretation of the original data. The processing of data leads to an inherent paradox that may not be readily apparent. The original unprocessed, or minimally processed, data are usually the most difficult to understand or use by anyone other than the expert primary user. With every successive level of processing, the data tend to become more understandable and often better documented for the nonexpert user. One might therefore assume that it is the most highly processed data products that have the greatest value for long-term preservation, because they are more easily understood by a broader spectrum of potential users. In fact, just the opposite is usually the case for observational data, for it is only with the original unprocessed data that it will be possible to recreate all other levels of processed data and data products. To do so, however, requires preservation of the necessary information about processing steps and ancillary data.

Another important characteristic of observational data is their volume. In this respect, observational data can be divided into two different classes: small-volume and large-volume data sets. The majority of traditional ground-based, in situ observations form small-volume data sets because they are based on individually conducted measurements or sample collections. Satellite and other remotely sensed observations generally form large-volume data sets.

The committee defines small-volume data sets as those with volumes that are small in relation to the capacity of low-cost, widely available storage media and related hardware. The hardware and software to write and produce CD-ROMs are now generally available for less than $10,000, and personal computers capable of reading CD-ROMs are being marketed as home-use, consumer items. For example, the total volume of the small-volume oceanographic data is projected to be less than 50 gigabytes by 1995, and thus the entire historical data set for all observations could be stored on fewer than 100 CD-ROMs. This is fewer diskettes than many people have in their compact disk music collections.

Issues such as archiving cost, longevity of media, and maintenance of the data holdings are not the dominant considerations with regard to retaining small-volume data sets. Rather, the major issue with respect to this class of data is the completeness of the descriptive information, or metadata. If a data set has been properly prepared and documented, the operations required to migrate the data should be amenable to significant automation and therefore pose only a minor challenge to the long-term maintenance of the archive. Further, these data may be widely distributed with simple replication of the media. For example, the various NOAA and NASA data centers have provided copies of their data sets to many users for a number of years.

A different problem is posed by large-volume data sets. The biggest data sets typically come from Earth observation satellite sensors and space science missions, and are challenging to some contemporary storage devices. However, it is clear that for the data set to exist at all, an adequate storage medium capable of capturing and maintaining the data for some time period must exist when the data are generated. Further, the time period for reliable, initial storage should at least cover the lifetime of the data set at the organization acquiring and using the data before the records need to be migrated to new media or transferred to another organization, such as NOAA or NARA. In addition, during the initial storage period, there are likely to be major increases in the density of mass storage accompanied by significant decreases in the cost of storage of the data. Thus, data sets that are challenging today will gradually be transformed to "small-volume" status in the future, as advancing technology increases the capacity and lowers the cost of storage devices. Nevertheless, it is important to note that the largest data sets (e.g., larger that one terabyte) can present significant organizational and management problems that require special analysis of the data flow, volume, access, and timing characteristics.

Observational Data in the Space and Earth Sciences

Astronomy and Astrophysics Data

Astronomy and astrophysics are observational sciences; that is, they are based on what the sky provides and we collect. Therefore, in many astronomical investigations there is no such thing as "repeating an experiment" with the expectation of getting the same results. Many objects have properties that change with time either because of their intrinsic nature (e.g., variable stars), evolution (e.g., stars going supernova), or reasons yet unknown. It happens quite frequently that a highly variable object is found in satellite data and subsequent archival research in optical plates allows its identification as a given type of star.

Astronomy and astrophysics data are acquired by both ground-based and space-based observatories. Ground-based observatories, which are operated by universities or other nonprofit organizations (e.g., Association of Universities for Research in Astronomy, the Smithsonian Institution) and funded by these organizations or by the National Science Foundation (NSF), have traditionally been used to study the sky at visible wavelengths. Since the second World War, astronomers have used improving technologies to observe at radio and infrared wavelengths. Consortia of universities, including both U.S. and foreign institutions, are constructing new telescopes, which use advanced technology to build larger mirrors that will allow us to look deeper into the universe. Radio observatories range from smaller ones operated by universities to larger national facilities, such as the National Radio Astronomy Observatory, funded by

NSF. Most telescopes are for individual observing programs, but some are dedicated to systematic sky surveys.

Data from ground observations have traditionally been the property of the observer; therefore, observatories have no standard policies for data archiving. The exceptions are some big projects, such as the Palomar Sky Survey, where data either are made public and sold or are archived within the university or observatory. Some centers, such as the National Radio Astronomy Observatory, the National Optical Astronomy Observatories, and the Harvard-Smithsonian Center for Astrophysics, have begun to archive most data obtained from major telescopes. These data are valued and used broadly by astronomers. Nevertheless, archival activities remain of generally low priority.

Although the older astronomical data consist of photographic plates and other analog data, virtually all data today are collected digitally. There also have been major efforts to digitize old photographic data to allow their analysis by computer. An example of this is the digitization of a whole-sky survey by the Space Telescope Science Institute, and this survey is now available for sale on CD-ROM from the Astronomical Society of the Pacific. Recently, the astronomical community adopted a standard format for transfers of digital files (FITS). With the advent of digital data, there also has been an evolution from individual data analysis packages to a few widely distributed packages (e.g., IRAF, AIPS, VISTA, XANADU), which provide standard tools for baseline analysis.

Because of the filtering and distortion produced by the Earth's atmosphere, the amount of energy emitted by celestial bodies that can be detected on the ground is limited significantly. Observations from space above the atmosphere remove such limitations. From its inception, space astronomy and astrophysics have been mostly under NASA's purview, although some important experiments have been financed by the Department of Defense. The data are collected through telescopes and detectors placed on airborne devices (balloons or planes), rockets, NASA's Space Shuttle, and orbiting satellites. The largest volume of data is collected by satellites, and most of these missions are international collaborations. The U.S. portion has always been handled by NASA.

Within NASA, space astronomy and astrophysics are organized in different wavelength-based disciplines, reflecting the organization in the scientific community. These disciplines include the infrared, whose main data center is the Infrared Processing and Analysis Center in Pasadena, California, where the data from the Infrared Astronomy Satellite mission are archived; the optical and ultraviolet, with data centers at the Space Telescope Science Institute in Baltimore, Maryland, where the Hubble Space Telescope data are archived, and at the NASA Goddard Space Flight Center in Greenbelt, Maryland, where the International Ultraviolet Explorer archive resides; and high-energy astrophysics, which maintains x-ray data at the Einstein Observatory Data Center in Cambridge, Massachusetts.

Table 2.1 provides a representative sample of NASA Astrophysics Archives. The earlier NASA astrophysics projects were so-called "principal investigator" missions, where a contract was awarded to a group of principal investigators, who built the hardware, received the data from the experiments, and analyzed and interpreted them. These principal investigators had no clearly stated guidelines to prepare data for archiving, other than to deliver the reduced data to the NASA data depository at the National Space Science Data Center (NSSDC) at the NASA Goddard Space Flight Center. Documentation generally was minimal, and the data, which often were not well-documented or well-organized, were difficult to retrieve for scientific use, even if they were adequately physically preserved.

It has become fully apparent, however, that the uniqueness and high acquisition cost of these space data make their effective preservation and archiving a high priority. Even after the active operation of a space observatory has ended, the data typically are retrieved and used by scientists for many more years. As a result, the situation has improved considerably at the NSSDC in recent years. Moreover, NASA now funds wavelength-specific scientific data centers to process the data, eliminate anomalies in the data, and provide software for scientific analysis.

TABLE 2.1 A Representative Sample of NASA Astrophysics Archives, by Satellite Mission

	High Energy Astrophysical Observatory 2	International Ultraviolet Explorer	Infrared Astronomical Satellite	Hubble Space Telescope	Compton Gamma Ray Observatory
Data type	X-ray data	Ultraviolet data	Infrared data	Optical/Ultraviolet data	Gamma-ray data
Year of launch	1978	1978	1983	1990	1990
Duration	2.5 years	Ongoing	300 days	Ongoing	Ongoing
Total data volume (gigabytes)	~100	~100	~150	~5500 by year 2005	~1000 by year 2000
Data center	Einstein Observatory Data Center, Cambridge, Massachusetts	National Space Science Data Center, Greenbelt, Maryland	Infrared Processing and Analysis Center, Pasadena, California	Space Telescope Science Institute, Baltimore, Maryland	National Space Science Data Center, Greenbelt, Maryland

Planetary Science Data

Planetary data also are acquired by both ground-based and space-based observations. Planetary data include observations of the entire physical system and forces affecting a planet or other body, including the geology and geophysics, atmosphere, rings, and fields. The sensors used collect data across much of the electromagnetic spectrum. Currently, most planetary observations are supported by NASA, either as the direct result of planetary missions or as ground-based observations that support a mission. Over the past three decades, NASA has sent robotic spacecraft to every planet in the solar system except Pluto, to two asteroids, and to a comet. Men have walked on the Moon, performed experiments there, and returned samples. The knowledge we have about the bodies in the solar system, with the exception of our own planet, comes mostly from space missions. In some cases, such as the gas giants Jupiter, Saturn, Uranus, and Neptune, robotic space probes have provided most of our current knowledge. Many of the satellites of the other planets were no more than points of light with minimal spectral and light-curve measurements before the Voyager mission. Now each is recognized as a separate world with highly individual characteristics.

The scientific and historical importance of space-based planetary observations, the realization that additional missions cannot replicate the original observations, and the expense of planetary missions all prompted NASA to create the Planetary Data System (PDS) to improve the acquisition, archiving, and distribution of planetary data. The developers and current staff of the PDS recognize that the data from planetary missions make up the scientific capital of the agency's planetary exploration program and that these data are a national resource. The PDS tries to acquire all existing planetary data from NASA's missions and even from international ventures, in order to have a complete archive of our exploration of the solar system. In addition to the space-based measurements, the PDS accepts relevant ground-based observations and laboratory measurements that support planetary missions by providing baseline or calibration data. A basic condition for acceptance is that the data set must be properly documented and include all relevant ancillary data, including planet and spacecraft ephemerides, calibration tables, and experimenter notes about the shortcomings of the data. Members of the PDS scientific staff and scientists in the community who have expertise within the relevant disciplines peer-review each data set.

One of the more important contributions of the PDS, especially with regard to the ongoing preservation of data in a useful form, is the electronic "publication" of the majority of the data from many planetary missions in the form of CD-ROMs. These include not only the data, but also documentation, format specifications, ancillary data, and even, in some cases, display and analysis tools.

Space Physics Data

Space physics involves the study of the largest structures in the solar system—the plasma environments of the planets and other bodies and the solar wind. Those environments consist of plasmas ranging from low energies (the thermal component) to charged particles of high energies, including cosmic rays accelerated by galactic processes. They also consist of the magnetic fields (if they exist) of planets or the Sun, as well as electrostatic and electromagnetic fields generated from natural instabilities in plasmas and charged-particle populations. Furthermore, in many locales, such as comets and the Earth's ionosphere, dust and neutral gases play an important role in mediating the behavior of plasmas and electromagnetic fields. As a consequence, the field of space physics requires a broad array of sensors and instruments at all levels of complexity.

Many instruments make in situ observations, but novel techniques enable remote sensing of various plasma regimes. Because some of the most apparent manifestations of space physics processes result in the northern lights and in planetary-scale modifications of the terrestrial magnetic field (and subsequent catastrophic effects on power grids and communications), space physics relies heavily on a wide array of ground-based observations, including magnetometers, ionospheric sounders, incoherent radar facilities, all-sky cameras, and photometers. In addition, a broad range of ground-based and space-based solar monitors has become crucial to study the correlations between various disruptions in the terrestrial plasma environment and solar activity, including sunspots, flares, and prominences.

For many reasons, it is essential to preserve space physics data for long periods of time. The Sun drives solar-terrestrial relationships, and many studies require observations over 22-year solar cycles. During this cycle the Sun reverses its magnetic polarity twice and goes through periods of increased activity with sunspots and associated flares. At solar activity minimum, flare and sunspot activity decreases, but expanded coronal holes appear. Long intervals of records are required because each solar cycle is different from previous ones and because there are long-term deviations, such as the Maunder minimum, from "normal" patterns. From the terrestrial point of view, there are motions of the magnetic dipole and even magnetic field reversals on time scales of thousands of years.

Because many space physics observations are taken in situ, models of the magnetosphere need data collected by many spacecraft, having different kinds of orbits and trajectories. To make sense out of data from one of these missions, it is important to be able to examine what another spacecraft in a different orbit found. Only by preserving the data from numerous missions do we acquire a sufficient archive.

Space physics has generated about 50 gigabytes of data per year over the last 30 years. The field has enjoyed this extraordinary productivity primarily because most missions were in Earth orbit and were tracked continuously for years. Many of these data sets were "archived" by sending the tapes—and sometimes the relevant documentation—to the NSSDC. Copies of the data on microfilm or on other media were sent there as well. Unfortunately, for every well-prepared, thoroughly documented space physics data set at the NSSDC, there are several poorly prepared and improperly documented data sets. For the earliest space missions, the archiving techniques were undeveloped, and archiving was not deemed a high priority. Thus, there are many data at the NSSDC that most scientists would find difficult to use with only the information originally supplied. Given the recent emphasis on the proper preservation of data and the importance of archiving—prompted in part by two General Accounting Office reports (1990a,b) and also by a heightened awareness and desire for high-quality archives by the community—many recently archived data sets are in better condition than their predecessors. Even though the Space Physics Data System has been in existence only since 1993, the more advanced data activities in other disciplines have influenced the space physics community favorably. Hence, it is becoming more likely that the data now being submitted are of a higher quality, have more adequate documentation, and are more complete than earlier data sets.

NOAA, NSF, the Department of Defense, private and educational institutions, and foreign organizations typically support the ground-based observations. Most of these data, not managed by NASA, eventually come under the purview of the National Geophysical Data Center, operated by NOAA at Boulder, Colorado. The center's holdings consist of over 300 digital and analog databases, some of which are very large. However, many important data sets still reside solely in the hands of the original investigators, the military, or foreign sources.

Atmospheric Science Data

Atmospheric science data sets are diverse and present a variety of problems for distribution, archiving, and later interpretation. Some data sets on the atmosphere stand out as the largest in any scientific discipline, particularly those from remote sensing by satellite or radar; others consist of contributions from thousands of individuals all over the world, and the provenance of those data is sometimes uncertain. Many data sets span decades, and a few span more than a century, with accompanying problems due to lack of homogeneity in measurement techniques and sampling strategies. The largest atmospheric science data holdings in the United States are those of the federal government. However, significant amounts of material are available only from state or private sources.

Not all atmospheric data sets are large and conspicuous; many are small. There are hundreds of data sets of only a few megabytes or less. There are also many medium-sized data sets that range from perhaps 100 megabytes to tens of gigabytes, as well as very large data sets, many terabytes in volume. Table 2.2 provides a sampling of some of the larger data sets. Data volume does not drive the cost of archiving small-sized and medium-sized data sets if proper technical choices are made. Rather, it is the labor-intensive process of readying a data set for indefinite preservation that can be costly.

Many atmospheric data sets are dynamic, continually growing or being otherwise modified. Because weather keeps occurring, observational time series from operational meteorological activities are never "complete." In contrast, field programs usually have finite extent, and the resulting data sets have a definite end. However, many recent large, complex field programs have spawned associated monitoring activities that have continued after the initial phases of the project. Despite the frequent usage of the term "experiment" to denote field programs, these intensive efforts are observational, rather than experimental, exercises. Some truly experimental data exist, including a few data sets that include the results from such work as sensor development and tests, fluid dynamics experiments, thermodynamic measurements, and laboratory chemical studies. Nevertheless, the vast majority of atmospheric science data describe observations of ever-changing phenomena, and thus they are unique, valuable, and irreplaceable.

For much meteorological and climate research, as well as for many applications, it is essential to have archives of global data. This goal has been largely achieved in the United States, although older data sets still need to be digitized. Collectively, U.S. archives have the best sets of global data of any nation, particularly for data since the early 1950s. However, many valuable data stored in other nations are inaccessible to U.S. scientists (and in some cases are inaccessible to those nations' scientists as well).

Meteorological and other atmospheric data are used for varying purposes on different time scales. It is convenient to delineate three: (1) real-time or current, (2) recent past or short-term retrospective, and (3) distant past or retrospective. Compared with other disciplines, meteorological data are probably used by a wider segment of the U.S. population than other scientific data, because they relate directly to practical, daily concerns. There is a large lay audience for weather and climate information.

The real-time or current use of most data sets usually motivates decisions on collection strategies and therefore quality. For example, the primary reason for collecting most meteorological data is for operational weather forecasting and warning, including forecasting for aviation operations. These data are perishable, and timeliness and spatial resolution are more important than absolute accuracy and continuity.

There are many recent past or short-term retrospective uses of meteorological data that can be of great significance. In this context, short term typically means from yesterday to a few weeks, or occasionally a few months, ago. A good example of such usage of data is in monitoring the development of a drought, a significant function for predicting crop yields. The transportation industry uses past data for verification of weather conditions for delay claims.

Most retrospective uses require data from several months old through the traditional (though now suspect) 30-year averaging periods used for climate normals. The National Climatic Data Center handles over 100,000 data requests per year. The state climatologists and regional climate centers also process about this many. Legal proceedings and insurance claims often require accurate meteorological records for corroboration of witness testimony, criminal investigations, and validations of weather claims related to accidents and property damage. Farmers and agronomists need data covering months to years for studies of pesticide residue and toxicology, decisions about pesticide spraying, planning of fertilizer usage, and crop selection. Architects and building engineers require site-specific data on heating and cooling needs, wind stresses, snow loads, and solar availability. Airport designers need prevailing wind patterns. Utility planners need aggregate heating and cooling loads for their areas.

Long-term retrospective uses of atmospheric data are the primary concern in this study. These uses are highly diverse, difficult to predict, and make great demands on the data and their associated metadata.

The Challenge: Preservation and Use of Scientific Data

TABLE 2.2 Volume of Selected Data Sets in Atmospheric Sciences

Type of Data Set	Comments	Dates	Years	Volume
Atmospheric In Situ Observations				
World upper air	Two times per day, 1,000 stations	1962-1993	32	25 GB
World land surface	Every 3 hours, 7,500 stations	1967-1993	27	60 GB
World ocean surface	Every 3 hours (~40,000 observations per day)	1854-1993	139	15 GB
World observations during First GARP Global Experiment	Surface and aloft, but not satellite	1978-1979	1	10 GB
U.S. surface	Daily, now 9,000 stations	1900-1993	94	15 GB
Selected Analyses (mostly global)				
Main National Meteorological Center analyses	Two times per day, increasing at 4 GB/year	1945-1993	48	50 GB
National Meteorological Center advanced analyses	Four times per day, increasing at 19 GB/year	1990-1993	4	58 GB
National Center for Atmospheric Research's ocean observations and analyses	Thirty-eight data sets			8 GB
European Center for Medium Range Weather Forecasting advanced analyses	Four times per day, increasing at 8 GB/year	1985-1993	9	76 GB
Selected Satellites				
NOAA geostationary satellites	Half-hour, visible and infrared	1978-1993	16	130 TB
NOAA polar orbiting satellites		1978-1993	15	
Sounders (TIROS Operational Vertical Sounder)			15	720 GB
Advanced Very High Resolution Radiometer (4-km coverage, 5 channel)			15	5 TB
NASA Earth Observing Satellite-AM	In development, 88 TB/year, level-1 data	1998-		
U.S. Radar Data				
Domains of 30 to 60 km		1973-1991	19	1 GB
Next Generation Radar System (NEXRAD)[a]	650 GB per radar each year, 104 TB/year for 160-site system	1997-		100s TB

Notes: Many other atmospheric data sets have volumes of only 1 to 500 MB.
1 MB (megabyte) = 10^6 bytes; 1 GB (gigabyte) = 10^9 bytes; 1 TB (terabyte) = 10^{12} bytes.

[a]First radars were deployed in 1993.

Most of the uses discussed above do not need data covering more than a few decades. Several of these applications, however, require the longest time series we can provide.

When technology advances and alters the method of data collection, there is a strong impetus to scrap the data collected by "obsolete" technology. However, these old data may become critical in the future. A notable example involves upper air wind profiles. These were originally collected by kites and later by radiosondes carried on balloons. With the onset of the space program, there was an urgent need for detailed low-altitude wind data for analysis of stresses on rockets at launch. Appropriate data could not be obtained from radiosondes, because of their high ascent rate, but older kite-based data, which had been scheduled for disposal, were available. Fortunately, they had not yet been destroyed when they were again needed.

There have been dramatic retrospective uses for military purposes (e.g., Jacobs, 1947). Planning for the D-day invasion of France, bombing runs over Japan, and the recent desert war in Iraq all required detailed climatic information, some long thought useless but not yet discarded. Such unexpected uses require the retention of many types of data from many places for a long time. Since the first flights of meteorological satellites in 1959, we already have had several examples of important retrospective uses of satellite data sets. For instance, a combination of reprocessed Nimbus-7 satellite data and old data from the Dobson network helped to confirm the recurring seasonal loss of stratospheric ozone over the Antarctic in the early 1980s.

If meteorologists are to study past weather events, such as severe hurricanes, damaging winter storms, or outbreaks of tornadoes, they must have at their disposal all data for the periods of time and geographical areas involved. Hurricane track records spanning more than a century are still regularly used for both research and operational purposes.

An increasingly significant use of meteorological data is the monitoring of the climate of the planet. Although barely two decades ago the study of climate was not a very high priority, today climate research issues are prominent; some of the nation's leading scientists specialize in climate studies, and policymakers seek information on likely climatic conditions of the future. The importance of old atmospheric data has become clear, but the reanalysis of these old data in the search for trends has often found them inadequate and poorly documented. The growing interest in global climate change and the difficulties with historical data that it helped uncover have strongly motivated earth scientists to take a serious interest in the long-term preservation of atmospheric data. Similarly, studies of long-term water and land usage require time series of many decades, or more. Such data needs also apply to planning aquifer usage and studies on deforestation and desertification.

Some historians examine connections between environmental conditions and human events. The time scales studied can range from the immediate, such as the influence of weather on battles, to the very long term, such as the rise or decline of a civilization affected by water availability. Workers in this field often search through the oldest existing data and have even provided meteorological information to atmospheric scientists from unconventional sources such as diaries and agricultural records.

Contemporary arrangements for the storage and archiving of atmospheric data are diverse, complex, and present many problems. Some of these arrangements could be improved. Atmospheric data are in many locations, and they have a broad range of life cycles. Difficult problems arise in preparing metadata, packaging data for extended archiving, motivating researchers to prepare their data for use by others, and simply dealing with the large size of some of the atmospheric data sets. Criteria for identifying data sets to save indefinitely are not necessarily obvious. Finally, any proposed solutions must be made in full recognition of their impact on budgets and other resources.

Geoscience Data

Spatially, the domain covered by the geosciences extends from the Earth's core to the surface and into space. Temporally, it covers broad trends from the remote origins of the Earth to possible future scenarios, but it also is concerned with rapidly varying, often short-lived phenomena. Data in the geosciences fall into two broad categories. One is the observation and description of unique events, such as earthquakes, volcanic eruptions, and floods. In most cases, such data need to be archived for a long time period, regardless of their quality. The other category consists of observations of quantities continuous in space and time, such as gravity and the Earth's magnetism and structure, seismic sampling, and groundwater distribution.

The volume of geoscience data obtained with public funding has increased dramatically over the past few decades. This increase is the result of several converging factors, including the extremely varied types of observational data collected by the scientific community; the large volumes available through better measurement techniques, more sophisticated instrumentation, and advancing computer technology; and increasing demand from not only the scientific community but also the general public, including engineers, lawyers, and statisticians. Nongovernmental and commercial institutions also are major collectors and sources of pertinent data.

Two examples—the Landsat database and the nation's holdings of seismic data—illustrate many of the characteristics and issues inherent in the long-term archiving of geoscience data. Other examples are provided in the working paper of the Geoscience Data Panel (NRC, 1995).

The Landsat database consists of multispectral images of the Earth's surface, which have been accumulating since the launch of Landsat 1 in July 1972. The archive includes digital tapes of multispectral image data in several formats, black-and-white film, and false-color composites of synoptic views of the Earth's surface, all from 700 km in space. This database thus constitutes an important record of the evolving characteristics of the Earth's land surface, including that of the United States, its territories, and possessions. The record documents not only the results of various federal government policies and programs, but also those of many state and local governments and private programs and activities. It further provides documentation of the impact of various large-scale episodic events, such as floods, storms, and volcanic eruptions, and is of great value to both current and future public and private activities.

Landsat data are currently available in either image or digital form from the Earth Resources Observing System (EROS) Data Center in Sioux Falls, South Dakota. The Landsat satellites were originally under the control of NASA. However, in 1980 they became the responsibility of NOAA. The currently operational Landsat 4 and 5 spacecraft were placed under control of the EOSAT Company in 1985. Under EOSAT's control, the data are not in the public domain, are significantly more expensive, and carry proprietary restrictions on their use. Beginning with the launch of Landsat 7, responsibility for the Landsat system will pass back to NASA, which will build and launch the satellite the late 1990s. NASA will operate the systems and deliver the data to the EROS Data Center for distribution. The data will once again be in the public domain, although the EROS Data Center still plans to charge more than the marginal cost of reproduction in fulfilling user requests. It is now widely recognized that the shift to private control of the Landsat system significantly reduced the access to and use of the data.

As of January 1993 the Landsat database contained more than 100,000 tapes of varying density and formats, and over 2,850,000 frames of hard copy imagery. Digital Landsat data are usually delivered to users as magnetic tapes. Other media, such as CD-ROMs and streaming tapes, also may soon be used. Data requests occur most frequently in reference to a particular geographic location, commonly expressed as latitude and longitude, for a particular time of the year, and meeting certain cloud cover limitations.

Landsat data are used widely across the spectrum of geoscience applications in both civilian and military operations and research. These include such applications as the impact of human activities on the environment, land-use planning and resource-allocation decisions, disaster assessment, measurement and assessment of renewable and nonrenewable resources, and many others. They are used also by the general public in any context where views of the Earth's surface are needed. Examples include such diverse applications as visual aids in elementary and secondary education, background for highway maps, and illustrations for magazine articles about various regions of the world.

The Landsat database is unique because data from any given area may be available at sampled instants over a period of more than 20 years, thus making possible for the first time the study of slowly varying phenomena on Earth. Even though data from the early 1970s may now have a low frequency of use, their potential value remains high and they represent a significant archival record.

In contrast to the Landsat database, seismic data are broadly distributed rather than concentrated in one data center or system. This example focuses primarily on seismic data from earthquakes and explosions, both nuclear and chemical. Some federal agencies, notably the U.S. Geological Survey (USGS) and NOAA's National Geophysical Data Center, collect and archive important seismic exploration data. In addition, the Department of Defense (DOD), Department of Energy (DOE), U.S. Nuclear Regulatory Commission (USNRC), USGS, and NOAA have been and continue to be engaged in the collection and archiving of earthquake and explosion data. These agency programs are carried out independently of one another with the result that each agency has its own data management and archiving policies and practices. Consequently, these data holdings are greatly distributed among the agencies in fundamentally different forms and formats.

Global earthquake data have been acquired systematically since the early 1960s, when the U.S. Coast and Geodetic Survey of the Department of Commerce deployed a global seismic network of about 130 stations called the World-Wide Standardized Seismographic Network (WWSSN) and produced an archive of photographic film "chips" of the 24-hour/day recordings at all stations. Researchers and other applications could obtain copies of these analog data at modest cost. The success of this precursor to today's global digital network cannot be overestimated, because the availability of a global data set in standard format from well-calibrated instruments permitted previously impossible studies of global seismicity patterns, earthquake source mechanisms, and the Earth's structure. These studies have led to a vastly improved understanding of the dynamics of the Earth as a whole, including tectonic plate movements, generation of new ocean floor, evolution of the Earth's crust, and occurrences of destructive earthquakes and volcanic eruptions.

The USNRC has funded the operation of regional seismic networks over much of the United States, some since the early 1970s, in support of programs for the siting and safety of nuclear power plants. USGS also has co-funded or separately funded regional networks for earthquake hazard assessments in seismogenic areas of the United States. However, changes in the funding priorities of USGS and USNRC in recent years have resulted in the interruption or discontinuation of some of these networks, particularly in the eastern United States. This has adversely affected data flow and seismic research. Seismic data have been archived in a broadly distributed, nonuniform mode by the organizations—mostly universities—that collected the data from the various networks. Many of these data have long-term value for characterizing in detail the tectonic activity of seismogenic areas in the United States.

In addition to the federal agencies, several private sector organizations now collect, distribute, and archive seismic data sets of long-term significance. The Incorporated Research Institutions for Seismology (IRIS), a not-for-profit consortium of universities and private research organizations, is engaged in a major development of a global digital seismic network of about 100 continuously recording stations (the Global Seismic Network) in cooperation with USGS. The project also includes a versatile, portable digital seismic array of up to 1,000 stations that can be deployed for various time intervals for special seismological studies. Data sets from the global and portable array are being permanently archived at the IRIS Data Management Center (DMC) in Seattle, Washington. The DMC also serves as the International Federation of Digital Networks' center for continuous digital data, which adds observations from many additional stations to the archive. IRIS funding for this activity comes primarily from NSF and DOD. Finally, individual universities, such as the California Institute of Technology, the University of California at Berkeley, the University of Alaska, the University of Washington, Columbia University, Memphis State University, and St. Louis University, also maintain archives of the seismic data that they collect.

The volume of digital data currently held and anticipated to be acquired by the IRIS DMC is summarized in Table 2.3. Although some data sets have been completed because they are project- or program-specific, most of the current operations continue to add large amounts of new data and implement new technology for recording, storage, retrieval, and distribution, thereby creating a dynamic, highly distributed archive whose holdings and access protocols change with time. For example, the IRIS

TABLE 2.3 Summary of Actual and Projected Data Volumes Archived in the IRIS Data Management Center

	Number of Instruments[a]	Projected Data Volumes (gigabytes/year)						
		1994	1995	1996	1997	1998	1999	2000
GSN	100	1,159	2,359	3,959	6,003	8,047	10,091	12,281
FDSN	146	370	670	1,070	1,530	2,050	2,670	3,416
JSP arrays	5	1,095	2,190	3,650	5,475	7,300	9,125	10,950
OSN	30	0	0	15	58	218	498	936
PASSCAL-BB	500	1,318	2,277	3,556	5,154	7,073	9,312	11,867
PASSCAL-RR	500	542	885	1,341	1,912	2,597	3,397	4,310
Regional-Trig	500	150	290	490	730	1,030	1,390	1,755
Total	1,781	4,634	8,671	14,081	20,862	28,315	36,483	45,515

Note: Abbreviations are as follows:
GSN   Global Seismic Network (IRIS)
FDSN   Federation of Digital Seismic Networks
JSP   Joint Seismic Program (with the former Soviet Union) (IRIS)
OSN   Ocean Seismic Network
PASSCAL-BB   Program for Array Studies of the Continental Lithosphere—Broadband (IRIS)
PASSCAL-RR   Program for Array Studies of the Continental Lithosphere—Regional Recordings (IRIS)
Regional-Trig Regional Triggered Recordings

[a]Projected numbers by year 2000.

Source: IRIS Data Management Center, private communication, 1994.

DMC recently began providing both archived and near-real-time data on the Internet, thereby greatly facilitating rapid access.

Significant volumes of exploratory seismic data obtained by geophysical contractors are held by the Department of Interior. These data are used by the federal government and by petroleum companies in preparing for oil and gas exploration activities. There are, however, various proprietary restrictions on access to these data by other users.

In summary, the sources of seismic data are diverse, the archiving is highly distributed, and the data are in many different formats with different metadata structures. Moreover, data sets with long-term scientific and historical value reside in both federal and nongovernmental organizations, although in most of the latter cases federal funds have paid at least in part for their acquisition, archiving, and distribution.

The users of seismic data are many and diverse as well. They include federal and state government agencies, universities, and private industry, particularly the petroleum industry. Thousands of individuals are direct or indirect users of seismic data. Certainly, the public as a whole is an end user of historical seismic data and information, including the location, magnitude, and damage associated with earthquakes around the world.

Most seismic data sets have been or are now used both for operational purposes and for research, although for operational activities the data are used primarily immediately following their collection. Examples of their use for operational activities include tsunami warning and the rapid determination of the magnitude, location, and fault mechanism of destructive earthquakes and their aftershocks, both to inform the public and to assist in emergency response and special monitoring. On a longer time scale the data are used for hazard reduction and seismic safety in seismogenic regions, including local zoning decisions for future development, and siting and safety of critical facilities such as nuclear power plants. Data are obtained and used for continuous global monitoring of earthquake activity and of threshold or comprehensive test bans on underground nuclear explosions. Of course, there also is a broad spectrum of research that uses historical seismic data, including studies of the physics of earthquake and explosive sources, propagation effects on seismic signals, imaging of the Earth's structures at all scales, seismicity patterns, and earthquake prediction or hazard estimation. Older data are important and are commonly used for most of these types of research. For example, establishing the recurrence rate for larger-magnitude earthquakes requires decades to centuries of observations, even in the most seismically active areas.

In conclusion, most of the seismic data have long-term value for scientific research, disaster mitigation, and various socioeconomic uses. The data are archived in a broadly distributed manner. However, only a fraction of the archived data are under the direct control of federal government agencies, and it appears that many of these data sets are not considered official federal records. Except for most commercial exploratory seismic data, federal funds have paid for much of the instrumentation, station operation and maintenance, collection, storage, and distribution of seismic data. These important seismic data sets should be kept indefinitely in a form accessible to both the scientific community and other users.

Ocean Science Data

The oceans and atmosphere are turbulent fluids, constantly changing over many spatial and temporal scales. The numerous types of data that describe the oceans are often unrelated to one another, and even those that are related frequently have nonlinear and poorly understood interactions. For example, temperature data from a specific point and time in the North Atlantic cannot be accurately predicted from data collected in the same place the year before, or even the week before, or from data collected at the same time 1,000 kilometers or even 100 kilometers away, or from salinity data collected at the same place and time. Each datum contributes unique information as long as it is accurate, corresponds to a different physical quantity, is obtained from a different time and place, and cannot be accurately computed from other existing data.

One source of oceanographic data is the field program. Large and small field programs conducted in support of specific research projects are the prime contributors of in situ and in vitro observational data sets for all the ocean disciplines. In situ data sets are those that are derived by processing the measurements from sensors immersed directly into the ocean environment. Processing of in situ data is largely automated, and so the data sets are relatively dense. In vitro data sets are produced by laboratory analyses of samples collected from the ocean environment. These laboratory analyses combine sophisticated measurement equipment with labor- and time- intensive procedures. Therefore, in vitro data are typically sparse. Remotely sensed observations also may be associated with field program data by synchronizing in situ sampling with the use of remote sensing platforms.

The harsh and remote nature of the world ocean environment has inhibited the establishment of a routine data collection system. Although several remote sensing platforms do provide daily monitoring of ocean surface conditions on a global basis, continuous measurement of subsurface conditions with adequate time and space resolution for effective monitoring is not a reality. The lack of continuous and comprehensive oceanographic data may contribute most to the inconsistent data management practices and lack of community-wide standards for data reporting and exchange in the ocean disciplines. Because of the need for daily global prediction, such standards and practices are much more highly developed in the atmospheric community. The establishment of the Global Ocean Observation System presents an opportunity to engage the ocean community in the identification and implementation of appropriate standards.

Like other observational data, oceanographic data extend beyond directly or remotely measured observations of the environment. The data products based on the analyses, interpretations, and presentations of aggregates of observations also must be considered in the design, implementation, and maintenance of any data management and archiving mechanism. The more traditional products, such as parameter grids and output from ocean models, will surely be supplemented from innovative sources likely to emerge from the interactive scientific collaboration and value-added services that are becoming increasingly available through electronic networks.

The principal federal agency ocean data holdings are at the NOAA National Oceanographic Data Center (NODC), the NASA Physical Oceanography Distributed Active Archive Center (PO.DAAC) at the Jet Propulsion Laboratory, and at several Navy centers, which hold mostly classified data sets. In addition, significant amounts of data are held by the universities.

Located in Washington, D.C., the NODC archives physical, chemical, and biological oceanographic data collected by other federal agencies, including data collected by principal investigators under grants from the National Science Foundation; state and local government agencies; universities and research institutions; and private industry. The center also obtains foreign data through bilateral exchanges with other nations and through the facilities of World Data Center A for Oceanography, which is operated by the NODC under the auspices of the National Academy of Sciences. The NODC provides a broad range of oceanographic data and information products and services to thousands of users worldwide, and increasingly, these data are being distributed on CD-ROMs and on the Internet. Table 2.4 presents a summary of the NODC's data holdings.

The PO.DAAC is a major federally sponsored oceanographic data center, which is operated by the California Institute of Technology's Jet Propulsion Laboratory in Pasadena, California. As one element of the NASA Earth Observing System Data and Information System, the mission of the PO.DAAC is to archive and distribute data on the physical state of the oceans. Unlike the data at the NODC, most of the data sets at the PO.DAAC are derived from satellite observations. Data products include sea-surface height, surface-wind vector, surface-wind stress vector, surface-wind speed, integrated water vapor, atmospheric liquid water, sea-surface temperature, sea-ice extent and concentration, heat flux, and in situ data that are related to the satellite data. The satellite missions that have produced these data include the NASA Ocean Topography Experiment (TOPEX/Poseidon, done in cooperation with France), Geos-3, Nimbus-7, and Seasat; the NOAA Polar-Orbiting Operational Environmental Satellite series; and the DOD's Geosat and Defense Meteorological Satellite Program.

SUMMARY OF MAJOR ISSUES

The results of scientific research are disseminated in this country through a hybrid system that includes professional society and other not-for-profit publishers, the commercial sector, and the government. The formal journals are published largely by the professional society and commercial sectors, while government agencies manage less formal reports (gray literature). Secondary services, such as abstracting and indexing, provide access to this literature, increasingly by electronic means. While there are strains in this system because of rising costs, increasing workload, and issues related to the protection of intellectual property, it has served U.S. science well and has been an invaluable link in the process of translating scientific advances into further advances, useful technology, and economic benefits.

The current system, however, is not well suited to handle the scientific electronic databases that are the focus of this study. The costs of maintaining these databases are typically too great to be covered by user fees; instead, these databases must be considered part of the national scientific heritage. Some government agencies have accepted responsibility for maintaining and disseminating data resulting from their own research and development. In some cases, this system is working reasonably well, but in others there are problems even with providing current access. Archiving for the long term raises questions in all cases, however.

A general problem common to all scientific disciplines is the low priority attached to data management and preservation. Experience indicates that new experiments tend to get much more attention than the handling of data from old ones, even though the payoff from optimal utilization of existing data may be greater. For instance, according to figures supplied by NOAA, NOAA's budget for its National Data Centers in FY 1980 was $24.6 million, and their total data volume was approximately one terabyte. In

TABLE 2.4 National Oceanographic Data Center Data Holdings (as of October 1994)

Discipline	Volume (megabytes)
Physical/Chemical Data	
Master data files	
Buoy data (wind/waves)	9,679
Currents	4,290
Ocean stations	1,645
Salinity/temperature/depth	1,557
BT temperature profiles	872
Sea level	125
Marine chemistry/marine pollutants	89
Other	68
Subtotal	18,325
Individual data sets, for example	
Geosat data sets	12,841
CoastWatch data	60,000
Levitus Ocean Atlas 1994 data sets	4,743
Other (estimated)	11,000
Subtotal	88,584
Total Physical/Chemical	106,909
Marine Biological Data	
Master data files	
Fish/shellfish	115
Benthic organisms	69
Intertidal/subtidal organisms	30
Plankton	32
Marine mammal sighting/census	21
Primary productivity	7
Subtotal	274
Individual data sets, for example	
Marine bird data sets	52
Marine mammal data sets	4
Marine pathology data sets	4
Other (estimated)	200
Subtotal	260
Total Biological	534
Total Data Holdings	107,443

Source: NOAA, private communication, 1994.

FY 1994, the budget was only $22.0 million (not adjusted for inflation), while the volume of their combined data holdings was about 220 terabytes! During this same period, the overall NOAA budget increased from $827.5 million to $1.86 billion.

With regard to laboratory data, government programs have existed since the 1960s to compile results from the world scientific literature, to check the data carefully, and to prepare databases of critically evaluated data. For instance, the National Institute of Standards and Technology operates its Standard Reference Data Program, which covers a broad range of data in physics, chemistry, and materials science. The Department of Energy also supports a number of data centers of this type. Despite chronic underfunding, these programs have produced databases of lasting value to the nation. To cite one example, the Mass Spectral Database managed by the National Institute of Standards and Technology, the National Institutes of Health, and the Environmental Protection Agency contains spectra of over 60,000 compounds. It has been installed in many thousands of mass spectrometers that are being used for monitoring environmental pollution, designing drugs, characterizing new materials, and many other applications. The government investment in creating and maintaining this database has been repaid many times over.

In the area of observational databases, the situation is mixed. Federal agencies collect large amounts of observational data, which in many cases are continuously added to the available record of Earth and space processes. The data sets resulting from these activities sometimes are well-documented and maintained in readily accessible form; but in many other cases, they are exceedingly difficult or impossible to access or use, and thus are effectively unavailable. In general, the agencies and other organizations do a good job of making data and information available to the scientists (primary users) during the active stages of projects and for some time afterward. Examples of notable successes include the NASA Planetary Data System, where the premise has been that the data have long-term value and must be accessible indefinitely into the future, and the NOAA National Data Centers, where the policy is to migrate archived data to new media every 10 years.

Technological advances have kept pace with the large growth in data volumes in scientific disciplines such that the long-term retention of all or nearly all of the data collected is feasible. Indeed, in most fields the entire collection of data from the past is not large in comparison with the current and anticipated data volumes that will be collected during only a year or two. However, significant fractions of the older data are difficult or in some cases impossible to access, because they have not been transferred to new storage media. This transfer often has received low priority because many data management and data retention activities are chronically underfunded and just handling the current data flow uses nearly all of the available resources. Thus, many valuable data sets are stored on low-density round tapes or on specialized magnetic tape media requiring hardware that is now obsolete or inoperable. For example, a large volume of the early Landsat coverage of the Earth resides on tapes that cannot be read by any existing hardware. Recent data-rescue efforts have been successful in getting older data into accessible form, but these efforts are time-consuming and costly. The reason these efforts have been undertaken, particularly in the observational sciences, is the recognition that retrospective data are vital to understanding long-term changes in natural phenomena. Given the extraordinarily rapid advances in computing and storage technology in recent years, planned periodic migration of data to new media will be increasingly important in all scientific disciplines to ensure long-term access to our scientific data resources.

It is axiomatic that a database has limited utility unless the auxiliary information required to understand and use it correctly—the metadata—is included in the record. An unambiguous description of the storage format is obviously essential for interpretation of an electronic database. The requirement is even more stringent to support meaningful access to data over the long term, because the hardware, software, and even the language by which formats are described will likely be different decades and centuries from now. The same is true regarding the scientific details of the content of the data. Auxiliary information such as environmental conditions (e.g., temperature and pressure), method of calibrating the instruments, and data analysis techniques must be given to be able to fully and correctly use the data. Providing this information is time consuming and costly if done retrospectively, but much less so if it is prepared at the time the data are collected. Documentation that is inadequate for understanding and using the data greatly diminishes the value of the data, particularly for secondary and tertiary users.

Another major problem inhibiting access to data is the lack of directories that describe what data sets exist, where they are located, and how users can access them. This, too, is especially a problem for potential secondary and tertiary users. In many cases the existence of the data is unknown outside the primary user groups, and even if known, there frequently is not enough information for a potential user to assess their relevance and usefulness. This realization has resulted in an interagency effort, led by NASA, to build a Master Directory of Global Change Data and Information. This Master Directory is intended to inform users of where data sets of potential interest reside and how to access them. Similar directories are needed in other scientific disciplines, as well as across all disciplines. The lack of adequate directories adversely affects the exploitation of our national data resources and commonly leads to unnecessary duplication of effort.

A significant fraction of the archived scientific data is held by the federal agencies that collected the data as part of their mission. However, a large amount of valuable scientific data gathered with federal funds is never archived or made accessible to anyone other than the original investigators, many of whom are not government employees. In many instances, the organizations and individuals that receive government contracts or grants for scientific investigations are under no obligation to retain the data collected, or to place them in a publicly accessible archive at the conclusion of the project. At best, scientists in the same field may be able to obtain desired data sets on an ad hoc basis by contacting the original investigators directly; secondary and tertiary users typically are unaware of the existence of the data and have no mechanism (other than personal contact) to access the data. Thus, data sets that commonly are gathered at great expense and effort are not broadly available and ultimately may be lost, squandering valuable scientific resources and much of the public investment spent in acquiring them. Clearly, there is a great need for the agencies to get more return on their investment in science by the simple expedient of making the data collected under their auspices accessible to others.

As seen from the discussion in earlier sections and addressed in detail in the individual discipline panel reports (NRC, 1995), there is a large and diverse collection of scientific data and information extant in federal agencies and nonfederal organizations, including state and local agencies, universities, not-for-profit institutions, and the private sector. At a minimum, those data that are acquired with the support of federal funding should be regarded as part of the National Scientific Information Resource.

Finally, NARA's holdings of scientific and technical data in electronic or any other form are very small in comparison to the data holdings of these other organizations. Moreover, NARA's budget for its Center for Electronic Records, which has formal responsibility for archiving all types of federal electronic records, was only $2.5 million in FY 1994, a budget lower than that of many of the individual agency data centers reviewed by the committee in this study. Given NARA's current and projected level of effort for archiving electronic scientific data, it is obvious that NARA will be unable to take custody of the vast majority of the scientific data sets that require archiving. Therefore, a coordinated effort involving NARA, other federal agencies, certain nonfederal entities, and the scientific community is needed to preserve the most valuable data and ensure that they will remain available in usable form indefinitely. The challenge is to develop data management and archiving infrastructure and procedures that can handle the rapid increases in the volumes of scientific data, and at the same time maintain older archived data in an easily accessible, usable form. An important part of this challenge is to persuade policymakers that scientific data and information are indeed a precious national resource that should be preserved and used broadly to advance science and to benefit society.

3
Retention Criteria and the Appraisal Process

The National Archives and Records Administration appraises and retains records on the basis of their informational and evidential value. It is concerned with records of long-term value—those records that will probably have value long after they cease to have immediate, or primary, uses. Although scientific databases can provide evidence of the research conducted by an agency, their value is primarily informational; it is based on the content of the records rather than on their description of activities by the agency that collected or created them.

Special problems arise in appraising scientific data for their long-term value, particularly beyond the community of research scientists working in the specific field to which the measurements refer. Scientific data are voluminous, constantly increasing, and often difficult for those in other fields to use in their original formats. The data typically are expensive to collect, provide baselines for future observations, enhance understanding of other data, and are of immense importance for advancing scientific knowledge and for educating new scientists. The data also are important to an understanding of the world in which we live; the data (or the conclusions drawn from them) may be important to economists, historians, statisticians, politicians, and the general public. At the same time, it is difficult to predict the full value of the data to researchers and other users decades or centuries from now, although past experience has shown that scientific data collected many years ago provide unique contributions to new understanding of our physical universe.

RETENTION CRITERIA

The criteria that follow are to be used during the appraisal process to determine retention of physical science data. They should be applied—by those responsible for stewardship—to all physical science data, whether created by small individual projects or in the course of large-scale research programs. Similar criteria and guidelines must be developed for data in other disciplines. This is a topic of primary concern not only to NARA, NOAA, and NASA, but to all scientists, data managers, and archivists who work with such records, and was provided in the charge to the committee as a central issue. Although the committee found that many retention criteria apply to both the observational and the laboratory sciences, significant differences are noted below. The metadata requirements, which tend to be either poorly understood or ignored, are given particular emphasis. Additional details and distinctions are discussed in the working papers of the discipline panels (NRC, 1995).

Criteria Common to Both Observational and Laboratory Sciences

- *Uniqueness of data.* Do other authenticated copies of the data under consideration already exist in an accessible repository that meets NARA standards of permanence and security? If so, are they adequately backed up? If the answers are yes, the data set need not necessarily be retained.
- *Accessibility—adequacy of documentation.* Though we might wish that all data sets were of high quality and accompanied by detailed metadata, that is not always the case. At a minimum, the metadata should be sufficient for a scientist working in the discipline to make use of the data set. If documentation is lacking or is so poor that a data set is not likely to be of value to someone interested in data of that type, or the data are more likely to mislead than to inform, that data set should have a low priority for archiving, or perhaps should not be archived even if resources are available. Nevertheless, the committee does not believe that many data sets should be purged because they lack sufficient documentation. The vast majority of data sets now meet minimum standards of documentation, which means that a skilled user either is given sufficient information or can figure it out. Adequacy of documentation is thus but one criterion to consider in the appraisal of data for long-term retention. Metadata requirements are discussed in greater detail below.
- *Accessibility—availability of hardware.* Is the hardware needed to access the data obsolescent, inoperable, or otherwise unavailable? If so, the data are not usable. Decisions on whether to keep such data should be based on the feasibility of building or acquiring the necessary hardware, the usability of the data if they were accessible, and the nature of the data set, if known. To avoid this situation, migration of data to current storage media should be part of the normal routine to maintain the archive.
- *Cost of replacement.* Could the data be reacquired if a future national need for the data were to arise? If so, would reacquisition of the data be more costly than their preservation? For the observational sciences, the answer is almost always that the data cannot be reacquired. The exception is with a data set in a discipline in which the changes of nature are so slow that the data could be recaptured at another time. For example, data on the fossil record of evolution contained in stratigraphic rock units could be reacquired.

The laboratory sciences generate data that can, in principle, be reacquired. The question is whether the data can be reproduced at an acceptable cost. Data sets in the laboratory sciences that are candidates for long-term preservation can be classified into three generic types: (1) massive records and data from an original experiment, particularly a costly "mega-experiment," that there is no realistic chance of replicating (e.g., data obtained from expensive facilities such as plasma fusion devices, or data of interest in physics and chemistry derived from special events such as nuclear tests); (2) unique, perhaps sample-dependent or environment-dependent, engineering data, many of which never reach the published literature; and (3) critically evaluated compilations of data from a large number of original sources, together with the backup data and documentation on selection of recommended values, that represent tremendous accumulated effort.

- *Peer review.* Has the data set undergone a formal peer review to certify its integrity and completeness, or is there documented evidence of use of the data set in publications in peer-reviewed journals? Have expert users provided evidence that this data set is as described in the documentation? Formal review of data sets is not now common. It should be encouraged, however, especially in the observational sciences. A good model is the peer review system for NASA's Planetary Data System. In the laboratory sciences, the critically evaluated compilations of data referred to in Chapter 2 have undergone extensive peer review.

Differences Between the Observational and the Laboratory Sciences

Data derived from laboratory experiments, such as the hardness of steel produced in a particular melt, differ from data based on observations of transient natural phenomena, such as the records of the 1993 midwestern floods. Thus, they stimulate different questions related to data preservation issues. As has already been noted, one difference arises from the fact that transient natural phenomena are not reproducible; the fact that the resulting observational data are "snapshots in time" sometimes means that the data have historical or evidential value in addition to their informational value. Observational data sets that provide a continuous time-series record of the physical universe, or of human impact upon it, are important to future generations for comparison and the identification of trends. In addition, many observational data sets represent major engineering or worker-intensive collection activities that warrant documentation and could not feasibly be carried out again.

Experimenters have good reason to believe that if and when their data are recreated in the future, instruments will be better. In many experiments, raw data (e.g., the initial sensor readings before any transformations, conversions, averaging, or corrections are made) may exist only for a fleeting instant before they are discarded or further processed. Even when raw (level-0) data are acquired and saved, principal investigators frequently fail to provide appropriate documentation because they do not expect anyone else to use these data. Instead, the processed data sets are more likely to have adequate metadata and meet the committee's other criteria for retention.

Quite the opposite situation seems to prevail for the observational sciences, where many secondary scientific users feel they need to be able to get back to the level-0 data and are becoming more active in demanding that the collectors of the data provide adequate metadata.

Special Issues in the Retention of Observational Data

All observational data that are nonredundant, reliable, and usable by most primary users should be permanently maintained. This judgment is based on the committee's belief that advancing technologies and better data management practices make it possible to stay ahead of the growing data volumes, as discussed in Chapter 4. It also is likely that it will be more expensive to reappraise data sets than simply to keep them. If the committee is wrong on these two counts, it may be possible that the volume of the data can be reduced through sampling techniques and through intelligent selection of the data sets of highest priority, as explained below.

Data sampling issues arise in measurement systems and in considering archival strategies to provide ready user access. Even before a data manager faces archiving decisions, many sampling rate decisions already have been made. For example, in the atmospheric sciences, we could easily sample temperature sensors and wind gauges 100 times per minute, but that frequency is unnecessary for nearly all uses. In general, it is necessary to keep only data properly sampled in time and space; that is, the sampling interval must be such that the most-rapidly-varying component is not aliased. At least two samples per cycle are required according to the Sampling Theorem. Thus reduction of oversampled data to the minimum sampling rate needed, coupled with lossless data compression, can significantly reduce data volumes with no loss of scientific content. However, if the phenomena of interest are slowly varying, then more rapid fluctuations, which might have value for other purposes, can be filtered out and the data reduced to retain the desired data unaliased; this technique can further reduce the data volume at the expense of losing higher-frequency data. The archiving of only "representative" subsets of our largest data sets is often suggested, but the notion raises difficult issues in statistics, data management philosophy, and budgeting. In concept, there may be acceptable procedures for the long-term archiving of representative subsets of large data sets, but no effective methodology exists today to choose those that would satisfy the needs of future users.

An example of the approach to deciding which observational data sets to retain comes from the atmospheric sciences. In this field the value of a data set as part of a long time series is an important criterion for archiving decisions. The temperature record for a given year from a station operating over a century is much more valuable than a similar record from a nearby station with a shorter lifetime. Studies of climate change and other types of environmental change find long time series to be essential. For example, confirmation of the seasonal stratospheric ozone depletion over the Antarctic in the 1980s required reference back to the Dobson column ozone data from the first half of this century for comparative purposes. The U.S. Historical Climate Network data are a high priority for archiving because they represent a long time series of high-quality data, with excellent metadata; this combination of attributes of data of a common type makes the overall data set exceptionally valuable.

Metadata Issues

The committee has arrived at several related conclusions concerning the importance of documentation, or metadata, to the effective archiving of scientific data. These include the following:

- Effective archiving needs to begin whenever a decision to collect data is made.
- Originators of data should prepare them initially so they can be archived or passed on without significant additional processing.
- The greatest barrier to contemporary and future use of scientific data by other researchers, policymakers, educators, and the general public is lack of adequate documentation.
- A data set without metadata, or with metadata that do not support effective access and assessment of data lineage and quality, has little long-term use.
- For data sets of modest volume, the major problem is completeness of the metadata, rather than archiving cost, longevity of media, or maintenance of data holdings.
- Lack of effective policies, procedures, and technical infrastructure—rather than technology—is the primary constraint in establishing an effective metadata mechanism.

This suite of conclusions led the committee to recommend that "adequacy of documentation" be a critical evaluation criterion for data set retention. The following discussion illuminates the multiple perspectives of metadata, the essence of the problem, and important elements of any metadata solution.

Perspectives on Metadata

The term metadata often is used to denote "data about data," that is, the auxiliary information needed to use the actual data in a database properly and to avoid possible misinterpretation of those data. The term is used in many scientific disciplines, but not always with precisely the same meaning. Some comments on different types of metadata may be helpful.

The most basic class of metadata comprises the information that is essential to any use of the data. An obvious example is the units in which physical quantities are expressed. If units are not specified, the numbers are ambiguous; at best, the user must attempt to deduce the units by comparison with other data sources. In dealing with observational data, the coordinates and the coordinate system (spatial and temporal) obviously must be specified. Laboratory data are often sensitive functions of some environmental condition such as temperature or pressure. For example, the boiling point of a liquid varies with pressure, so that a boiling point value has no meaning unless the pressure is specified. Although this is well known, many mistakes occur when a user assumes a value taken from a compilation to be a boiling point at normal atmospheric pressure, while it actually refers to a reduced pressure.

A significant problem in planning a long-term data archive is simple carelessness on the part of the creators and custodians of the data. Current practitioners in a scientific field may implicitly understand what the units or environmental conditions are. Shortcuts are taken by the authors that cause no problem in communicating with their contemporary colleagues (although they may be confusing to those in a different discipline), but practices and language can change over a generation or two. For a long-term archive, even the most obvious metadata should be specified in detail.

Beyond this basic type of metadata, there is auxiliary information that is not needed by the majority of users (present or future), but is of interest to a few specialists. Included here are the parameters that have only a slight influence on the data in question, so that most users do not need to know about them. For example, the typical user of a database of atomic spectra is concerned only with the wavelength and a rough value of the intensity of each spectral line. However, a few users who are trying to extract further information from the data may want to know the conditions under which the spectrum was recorded, such as the current density, type of electrode, and gas pressure. Referring to the JANAF Thermochemical Tables, which are discussed in the Physics, Chemistry, and Materials Sciences Data Panel report (NRC, 1995), most users are perfectly content with the values given (along with the confidence that the compilers did a good job of selecting the most reliable values). A minority of users, however, will want more details on how the data were analyzed, such as whether the heat capacity values were fitted to a fifth-degree polynomial or a cubic spline, and so forth.

Perhaps the most pervasive form of metadata is the accuracy of the values. To a purist, no number has meaning unless it is accompanied by an estimate of uncertainty. Specifying the uncertainty of each data point increases the size and complexity of the database, but sometimes may be necessary. At a minimum, the metadata should include general comments on the maximum expected errors, even if a quantitative measure such as standard deviation cannot be given. Finally, the term metadata is sometimes understood to encompass the full documentation necessary to trace the pedigree on the database. For laboratory data, this includes citations to all the primary research papers relevant to the database. A critical evaluation of especially important quantities (such as the fundamental physical constants or key thermodynamic values) may end up with only a few hundred data points, but include massive documentation and citations to a hundred years of literature. In such cases the metadata occupy far more space than the data themselves.

From this discussion, it is evident that metadata can span the range from a few simple statements about the data to very extensive (and expensive) documentation. It is difficult to give general guidelines on the amount of metadata needed; each case must be considered in the context of how future users may use the data and what auxiliary information they will need. Some guidance may be obtained from formal efforts to set metadata standards for experimenters to follow in preserving their data. In chemistry, for example, many organizations have developed detailed recommendations on reporting data from specific subfields. These have been collected in a recent book, *Reporting Experimental Data* (ACS, 1993). The American Society for Testing and Materials Committee E49 on Computerization of Material Property Data has an ambitious program to develop consensus standards for metadata requirements for databases of properties of engineering materials. These documents emphasize that metadata requirements must be approached on a case-by-case basis and must involve experts in each field.

The conclusion is that metadata, whatever the particular form, are crucial to the use of almost every data set and must be included in any archiving plan. The necessary metadata usually add very little to the storage requirements, but may require considerable intellectual effort to prepare, especially if they are assembled retrospectively rather than when the data are first collected.

The preceding discussion defines metadata from the perspective of the research scientist. An additional, and somewhat overlapping, perspective is provided by the computer science community. In this community, the term metadata refers to the specification of electronic representation of individual data items, the logical structure of groups of data items, and the physical access and storage media and formats that hold the data. To the computer scientist or database administrator, the contextual data that the research scientist refers to as metadata encompass other data entities. In fact, divergence can exist even among research scientists as to the differences between data and metadata. What is metadata for one may be data for the other.

In view of this confusion, the committee has chosen to keep the term metadata and to explicitly define its fundamental components. As such, the committee views metadata as representing information that supports the effective use of data from creation through long-term use. It spans four ancillary realms: content, format or representation, structure, and context. The content realm identifies, defines, and describes primary data items including units, acceptable values, and so forth. The representation realm specifies the physical representation of each value domain, often technology dependent, and the physical storage structure of aggregated data items, often arbitrary. The structure realm defines the logical aggregation of items into a meaningful concept. The context realm typically supplies the lineage and quality assessment of the primary data. It includes all ancillary information associated with the collection, processing, and use of the primary data. On the basis of this explicit definition, the following section describes metadata objectives, implementation issues, and potential for defining a standardized framework.

Analysis of Metadata: From Challenge to Solution

The problem of data set documentation is receiving increased attention in the context of scientific data management. In the earth sciences, global climate change research and general environmental concerns have ignited interest in a more interdisciplinary and long-term approach to conducting science. Interdisciplinary collaboration requires more effective sharing of data and information among individual researchers, disciplines, programs, and institutions, all of which may operate under different paradigms or have different terminology for similar concepts (NRC, in press). Further, long-term research requires that researchers be able to access and compare data sets that were created by past researchers and collected in different contexts by different technologies. Therefore, to support the interdisciplinary sharing and long-term usefulness of data, adequate metadata must be included within a framework that accomplishes the following objectives:

- provides meaningful selection criteria for accessing pertinent data;
- supports the translation of logical concepts and terminology among communities;
- supports the exchange of data stored in differing physical formats; and
- enhances the assessment of data sets by consumers.

A critical question is how to motivate the user community to participate in the process of metadata preparation and standardization. The issue of motivation is best addressed by the value system of the community itself. It may be argued that the problem will not be solved until the production of verified data sets and their provision to scientific colleagues become more highly valued activities. Developments such as the peer-reviewed publication of data sets should contribute to this shift in values. However, until these activities are assimilated into the fabric of career advancement, such as being incorporated into criteria for tenure in academic institutions, progress will continue to be slow and uneven.

Nevertheless, there are a number of specific actions that can be taken to promote the preparation and standardization of metadata. Funding agencies could help facilitate change by requiring and enforcing minimal documentation of data sets created under their grants (as well as other desirable data management and archiving practices discussed elsewhere in this report). This will not be an effective mechanism, however, unless the minimal standards for consistency and completeness are provided as a target for grantees and as a measuring stick for the funding agent. To be effective, these standards must be created through the collaboration of researchers, data managers, librarians, archivists, and policymakers.

Individuals and institutions in the scientific community could contribute by recognizing that data management and the provision of appropriate documentation of data are an essential science infrastructure function spanning all disciplines. Greater cost-effectiveness, consistency, and quality can be achieved if the many diverse data management activities are better coordinated. The essential requirement for making these value system changes and developing effective solutions is the recognition that all segments of the scientific community need to be educated on this issue. Funding agencies and the scientific community thus must move forward together in the development of a coherent strategy for end-to-end management that focuses on metadata requirements as a major element.

The ultimate solution for metadata handling will include an approach that not only supports the documentation of a data set throughout its life cycle, but also supports evolutionary documentation requirements. For example, early in the development and use of an instrument system, the scientific community may not be able to specify completely what metadata will be important for the effective use of the observations produced by this system. In this case, some of the documentation may include free-form narratives without the benefit of controlled vocabularies. Documentation of this nature is useful only to a limited audience that understands the specialized vocabulary of the source instrument, project, discipline, or institution. In addition, it is still difficult to make these descriptions useful to an automated agent performing a search on behalf of a user. As instrument use becomes more routine, this documentation could evolve to a more structured, but not cumbersome, form. One potentially useful approach constrains the textual descriptions to a well-defined, controlled vocabulary. If the vocabulary is clearly specified and made easily available with the data and associated documentation, users beyond those closely associated with the creation of the data set may be able to use this information to assess its relevance, significance, and reliability. Eventually, this more structured alternative will evolve into the specification of structured records with appropriately defined fields, standard value domains, and relationships with data set records. The committee also expects that improvements in software for natural language understanding will enable the automatic translation of free-form narratives into easily searched metadata fields.

An equally important component of the metadata solution is the identification and detailed definition of classes of information that are critical to the complete and consistent documentation of data sets. Information modeling techniques can be used to develop these classes of information, some of which will have clear, concise definitions and a set of defined attributes, while others will be identified but will not have clearly defined attributes or boundaries with other classes. The resulting information model should present a technology-independent description of metadata entities and their relationships with the primary data. The model should identify metadata that may be generalized across all classifications of data sets and usage patterns, as well as accommodate specialized needs. Such a model should provide the basis for intelligent information policies, data management practices, and metadata standards. The information policies, however, must not saddle data providers with long, cumbersome "forms" to fill out. That would discourage the contribution of the data themselves, and the committee recognizes that data with incomplete documentation are better than no data at all. Nevertheless, appropriately established metadata standards do not necessarily need to be difficult or costly to apply, and therefore need not be onerous to the data provider. An example of a generalized metadata framework in the observational sciences is presented in the working paper of the Ocean Sciences Data Panel (NRC, 1995).

OTHER ELEMENTS OF THE APPRAISAL PROCESS

A data management plan should be created for any new research project or mission plan, consistent with the requirements of OMB (1994) Circular A-130. A good example of this is the Project Data Management Plan of the NASA National Space Science Data Center (NASA, 1992). At a minimum, those individuals who have responsibility for implementing the data management plan and ensuring accessibility and maintenance of the data should play a key role in the subsequent appraisal process.

Most individual investigators and peer reviewers do not recognize their roles as appraisers for archival purposes, but the views of these experts should weigh heavily in the decisions relating to long-term value or permanency of the data obtained. The principal investigators and project managers who collect and analyze the data clearly have the best sense of how long the data will be valuable for their own scientific purposes. Primary users also can provide a detailed understanding regarding the uses of the data for their own discipline, but they may not comprehend the long-term value of the data for application to other research or national problems. Because such primary users and other data collectors sometimes do not think beyond their own needs, the agencies should work with NARA to provide good documentation at the inception of scientific projects, especially documentation that would be useful to secondary and tertiary users. Although providing more extensive documentation often may be viewed as an extra burden by the principal investigators and data managers, the labor and expense can be minimized if it is planned at the inception of a project, whereas it is extremely difficult after the project is completed. Proper data management practices can be promoted by considering data management in the evaluation of an investigator's past performance.

Because many scientific endeavors require participation by a number of agencies and organizations, it is important to coordinate data management activities and assign responsibilities for the maintenance of the data during periods of primary use. NARA is currently responsible for the final appraisal of federal records and the determination of their value as accessions to the permanent national collection under its statutory mandate. However, NARA should take advantage of the expertise of the other participants involved throughout the life cycle of the data.

The committee believes that all stakeholders—scientists, research managers, information management professionals, archivists, and major user groups—should be represented in the broad, overarching decisions regarding each class of data. The appraisal of individual data sets, however, should be seen as an ongoing, informal process associated with the active research use of the data, and therefore should be performed by those most knowledgeable about the particular data—primarily the principal investigators and project managers. In some cases, they may need to involve an archivist or information resources manager to help with issues of long-term retention. Although the committee believes that formal appraisals should be kept to a minimum, appraisals should be performed according to the data management plan established for each project.

Although the committee was not expressly charged with advising on classified data, there is an obvious need to save classified scientific data as well. The complete records of the atmospheric atomic bomb tests are a clear example. It is more difficult to provide and assess metadata for a classified data set, and it costs more to maintain classified data. Also, there is a trade-off between the value of the data for national security, the risk to national security if the data are declassified, and the potential value to society of having the data declassified. Thus, it is highly beneficial and cost-effective to have mechanisms in place that consider these issues periodically for any given classified data set and that promote declassification when appropriate.

RECOMMENDATIONS

The committee makes the following recommendations regarding the retention criteria and appraisal process for physical science data:

As a general rule, all observational data that are nonredundant, useful, and documented well enough for most primary uses should be permanently maintained. Laboratory data sets are candidates for long-term preservation if there is no realistic chance of repeating the experiment, or if the cost and intellectual effort required to collect and validate the data were so great that the long-term retention is clearly justified. For both observational and experimental data, the following retention criteria should be used to determine whether a data set should be saved: uniqueness, adequacy of documentation (metadata), availability of hardware to read the data records, cost of replacement, and evaluation by peer review. Complete metadata should define the content, format or representation, structure, and context of a data set.

The appraisal process must apply the established criteria while allowing for the evolution of criteria and priorities, and be able to respond to special events, such as when the survival of data

sets is threatened. All stakeholders—scientists, research managers, information management professionals, archivists, and major user groups—should be represented in the broad, overarching decisions regarding each class of data. The appraisal of individual data sets, however, should be performed by those most knowledgeable about the particular data—primarily the principal investigators and project managers. In some cases, they may need to involve an archivist or information resources professional to assist with issues of long-term retention.

Classified data must be evaluated according to the same retention criteria as unclassified data in anticipation of their long-term value when eventually declassified. Evaluation of the utility of classified data for unclassified uses needs to be done by stakeholders with the requisite clearances to access such data.

4
The Opportunities: The Relationship of Technological Advances to New Data Use and Retention Strategies

Rapid progress in information technology continually alters both the quantity and the quality of scientific information and periodically stimulates fundamental modification of data management and archiving strategies. Recent technological advances have enabled new methods and strategies for data storage and retrieval and have created better ways of connecting users to data resources and to each other. Moreover, the evolving technologies are catalysts for revising organizational structures to manage scientific data archives much more effectively in a distributed manner. Assumptions about effective management of scientific data that have been long and firmly held are being directly challenged by new information technology. These assumptions have been based on experience with management of paper records, generally in domains outside of science. Some of the outdated assumptions that are rapidly losing their relevance include the following:

- *Physical possession of the data is essential to their management and archiving.* This principle has outlived its usefulness in the context of electronic physical science data and has made access difficult for legitimate users. Electronic information is easily copied and disseminated. This feature removes constraints imposed by the limited physical access. Because most government physical science data are considered to be in the public domain, the constraints of copyright and fee collection to the free movement of data are removed as well.
- *Cost of an archive increases in proportion to collection size and use.* Physical archive cost is a function of space, as well as cataloging, repair, and access efforts. Improved inventory technology has eased some of the cost burden over the last several years, but, fundamentally, archives with large physical holdings operate in traditional ways with linearly scaling costs. Such costs actually discourage use, since physical handling of items scales with use, whereas budgets reflect usage indirectly. In contrast, electronic information storage and management costs have declined as rapidly as the costs of computer technology and processing over the last 30 years. There is no foreseeable end to this process. Storing and using the next byte will be cheaper than storing and using the most recent byte for a long time to come.
- *Only archivists and librarians have the capabilities to manage archived data.* While librarians and archivists are important advisors and participants in scientific data management, the dominant management responsibility falls to the scientific community and its designated scientific data managers (who are a blend of scientist, computer scientist, and librarian/archivist). If practicing scientists do not participate in the management of scientific information, such data will fall into obscurity or obsolescence.

- *The locator information (catalog) about the managed objects is simple and compact.* Finding relevant scientific information often requires searching the full content—and this content generally is not in the conveniently compressed form of text. For example, to search for all data sets where the stratospheric ozone concentration is less than some ad hoc threshold in some region, one would need to execute a complex algorithm on every data sample covering the region in question. Queries such as this become even more complex if the region of interest is determined after retrieval (e.g., how many days in a row was the areal extent of the ozone hole over open ocean greater than 5,000 square kilometers?). The selection and use of scientific data to solve complex problems can be simplified through the use of the concept of browsing information based on content. Browsing often involves examination of large numbers of samples and data volumes. Specialized "browsing products" can be defined to locate records of interest. For the query examples above, low-resolution ozone maps could be used to find candidate data sets with high probability of relevance. Information about the processes (including sensor characteristics, computer program capabilities, and calibration points) used to develop the data set is needed for its proper use. Such information increases the size and complexity of the locator service.

The remainder of this chapter describes how advancing information technologies enable the data manager, librarian, and archivist to deal with the challenges of scientific data management—in a collaborative fashion with the scientific user community.

ENABLING TECHNOLOGIES AND RELATED DEVELOPMENTS

Table 4.1 provides a summary of aspects of scientific data management changed by new technologies and related developments. These six areas are discussed in more detail below.

High-Performance Computer Networks

The rapid expansion of computer networks and their use for electronic mail and database access have obviated the need for researchers and other users of scientific and technical data to be in physical proximity to colleagues, information resources, and even advanced technical facilities. This has presented a menu of choices about the best means to distribute data and the responsibility of managing them.

A worldwide, "virtual" library is being created on the Internet. Application programs such as Mosaic are demonstrating the power of free and simple navigation across an ocean of available resources. Improving network capacity, reliability, performance, and security measures are helping to make these resources more widely accessible and useful.

High-performance networks also support movement of information for new applications (e.g., for producing safely managed backup copies, "profiling" information for individual user's needs, or staging data through a number of refinement steps in different locations for focused research). Networks support collaborative work and research projects that span traditional research boundaries. Such work requires easy access to a variety of data sources at once.

High-performance networks enable scientific data resources to be widely distributed and managed by groups of scientists. Users thus are freed to concentrate on the most effective use of the data, rather than on their own data management issues. Networks can provide a vehicle for regularly distributing backup copies of data and metadata to ensure safe storage. Distribution of data to users can be done via the network in addition to, or instead of, via physical media such as tapes and CD-ROMs. Data can be linked together to help users navigate among related items. This kind of linking is at the heart of the World Wide Web concept and brought to users by Mosaic. The population of information providers (e.g., people who can contribute to the knowledge base) has now grown to include all networked members of a user

TABLE 4.1 New Technologies and Related Developments That Enable a New Strategy for the Management of Scientific and Technical Data

New Technology Trends and Related Developments	Key Features	What Is Enabled?
High-performance computer networks	Distributed functions; rapid delivery of large data volumes	Location of databases and archives where best managed; collaborative work; distributed organizations; distributed responsibility
Low and declining cost of storage	Inexpensive backup; continually declining cost; ease of migration	Deferral of archiving decisions; trust in distributed management due to safe storage backup
Advanced data management	Ability to rigorously and formally manage diverse data types	More complex data structures (other than "flat files") handled in archives, with great potential advantages
Changing requirements for information technology professionals	Ability of personnel with lower technical skills to succeed in data management roles	Ability to entrust scientific data management in a distributed environment
High reliability of technology components	Availability of better components and connections; reduced procurement and operations costs	Reduced cost and effort in data migration; trusted connections for communication and collaboration
Development and acceptance of standards	Agreement on terms, interfaces, media, procedures	Reduced effort to communicate and apply results of others; ability to concentrate on mission issues and not on technology support

population. Such contributions can be as simple as an annotation on an existing item, or as complex as a fully processed and peer-reviewed new item. Most profoundly, the evolving network infrastructure enables new concepts for distribution of functions and responsibility in organizations (NRC, 1994).

Although networks can provide a quick and easy means to distribute data, it must be noted that CD-ROMs have been used to distribute data for several years and have been very successful. CD-ROMs not only permit users to have a huge local library of data, but they often come with a better set of data access tools than are normally available. Some data sets are large enough that the most cost-effective method to deliver them is on media such as Exabyte tapes (8 mm).

Low and Declining Cost of Storage

As for most aspects of computer hardware, the cost of storage has declined continuously and rapidly for the 30 years of the modern computer age. New storage technology is also increasingly compact and supports ever greater access speeds (Gelsinger et al., 1989). The historical trends are expected to continue for up to 20 years. Already, laboratory engineering results confirm this projection for at least the next decade. The most significant implication is that the decisions about sampling or discarding scientific data can generally be deferred, particularly for data sets for which the necessary metadata exist and whose quality has been certified. For relatively smaller data sets, the deliberation regarding long-term retention may well cost more than the recurring acts of migration. The cost of storage is small in relation to overall mission or investigation costs and therefore should not be a decision driver. Experience suggests, however, that the funds to meet these costs need to receive special protection in the annual agency budget cycles. The support for the data management aspects of scientific missions has typically had a lower priority than the data collection aspects. The low cost of storage also implies that the incremental cost of supporting a remote safe copy of data and metadata also will be small, except for the very large data sets. Therefore, over the next few decades, data received and stored may be expected to be cheaply and quickly migrated to new technologies when storage media reach their nominal limits of reliability or for convenience of improved access.

It is important not to expect a perpetual advantage from this technological discontinuity. The fact that data require significant time periods for their migration must be considered. The cost decay trend will slow down at some point in the future, causing the overall cost of storage to return to something closer to the linear relationship to volume. We also must be realistic and expect that funds will not always be available to save and back up every data set. Decisions on retention or sampling will have to be made.

Nevertheless, the already low and continually declining cost of storage allows a priori decisions to be made in certain circumstances to keep scientific data sets indefinitely. Backup or safe storage copies of data are becoming more affordable as data migration becomes less expensive with smaller, faster, and cheaper storage devices. Reliability also is improving with new software-based archive systems (including migration and backup features). However, there is an enhanced need for ongoing technology monitoring by an appropriate body for media, standards, and migration automation. Such monitoring should be incorporated in any scientific data management and archiving strategy.

The rapid change of storage technologies suggests that efforts to protect today's scientific data legacy must be accelerated. The obsolescence of media types and recorders/players is occurring within shorter and shorter time periods. This implies that "salvage" activities will be increasingly difficult for data left out of migrations to new media. This "join or be left behind" by-product of rapid technological change intensifies short-term budget pressures on archives. It demands in response a strong management commitment to provide resources and save important data sets.

If digital data are to survive, it is of fundamental importance to manage and constrain the costs of archive maintenance. The problem is that new data will be coming in, old data will need to be migrated to new media, the building will need to be repaired, and there usually will not be a lot of extra money for new equipment or added staff. To avoid problems, the data migration process in the system design must be almost totally automated. This refinement often has not been achieved, and it can cause unnecessary budget difficulties. Finally, it is essential for agencies to preserve all the hardware and software necessary to access all their data until the data have been successfully migrated or otherwise disposed of.

Advanced Data Management

There are signs that data management technology is beginning to address and, perhaps, to catch up with the complexities of the very large volumes of scientific data. Improvements have occurred in database management systems, hierarchical file systems, data representation standards, query optimizers, data distribution techniques, specialized access methods, and data security tools (Silberschatz et al., 1991). Further, investment in standards and cooperative approaches is accelerating, fueled in part by the demands of medicine, education, entertainment, journalism, financial services, and other commercial applications. While competing approaches and inconsistent vocabulary create near-term confusion, the attention and investment levels bode well for the longer-term capability to go beyond "flat file" representations of data that need to be archived. The new tools and techniques are more descriptive of the data, their heritage, the processes that have worked upon the data, and the relationships of data to each other.

New data management technology will enable easier representation of more diverse types of scientific data. Because of the rigor that new techniques require (e.g., for self-documentation or for precise definition of access methods), long-term archives will benefit from data structures other than flat files. The new technology also implies that the creation of a richer set of metadata will be easier to implement and that these data will be of high scientific value for content-based retrievals. To realize the potential of this enabled facility with metadata, the scientific community will have to accept and support efforts to develop and apply new metadata requirements.

The Changing Requirements for Information Technology Professionals

Information technology professionals with high skill levels can now be found in all parts of the United States and around the world. But as they bring the information technology industry to higher levels of maturity, the effect is to reduce the complexity of major tasks in managing information. Such tasks previously required their skilled use of sophisticated assembly language or job control language (JCL) programming. JCL programming refers to the steps in the old days that one used at the system console to get programs to run, attach the right files, print to the right printer, and similar functions. Today, much of this work is masked, made automatic, and controlled through icons and other means. These tasks can now be performed by competent scientists or professionals with lower technical skills, rather than by highly trained specialists. Because more functions can be completely handled by machines, management of the data can be greatly automated and operated by less skilled individuals. The data themselves can be widely distributed without fear of loss, particularly with a backup copy in safe storage.

Over the next 5 to 10 years, the costs for information technology professionals at individual scientific data centers and archives can be dramatically reduced. The reasons for the reduction in costs include more automatic processes for storage management, rudimentary learning capability in systems, services performed by end users based on their preferences, improved systems management, higher component reliability, improved application of standards, and vendor consistency with standards.

Although the dominant trend will be for a smaller, less technically skilled staff to manage the physical aspects of the archive, there will be a pressing demand for fewer, highly skilled people who blend the skills of physical scientist, computer scientist, and archivist. These people must be able to handle the intellectual challenges of bridging these disciplines while providing the coaching and direction to help develop data and operations standards for scientific communities.

High Reliability of Technology Components

Microprocessors, new storage media technologies, mature software, error correction capabilities, improved packaging, and reduced power consumption have all made significant contributions to the reliability of computer systems and networks. What was recently considered unreliable, requiring constant attention and expensive repair, is now regarded as reliable and not worthy of effort to repair. Although precautions have always been taken to protect against loss of valuable data, many of these precautions are now built into the base of mature software or are increasingly familiar parts of facilities' operating procedures.

High reliability of technology supports a capacity for high levels of trust and the ability to widely distribute functions and databases. These distributed systems can achieve the same levels of quality and trust as centralized archives through the use of the same underlying hardware and software technology, operating procedures, safe storage of copies, and high-quality (error-corrected) telecommunication connections. High reliability has enabled new applications such as the World Wide Web, in which context switching from one machine to the next—on a worldwide basis—is readily accomplished. Increased reliability also has allowed computing technology to be put into the hands of business managers, consumers, and shop clerks. Without such reliability, maintenance effort would outweigh productivity benefit. As a result, powerful organizational or operational frameworks can be built, much as new materials enable new architecture or new machines.

Development and Acceptance of Standards

The development of effective standards has been pivotal to promoting the widespread use of electronic information. Communication protocols such as TCP/IP have fueled the growth of the Internet. Other format standards for documents support their interchange. For example, the Standard Generalized Markup Language (SGML) provides a uniform way of formatting textual documents so that they can be read by different document processing tools. The HyperText Markup Language (HTML) is a standard used to represent and link documents; it is used to describe pages viewed with Internet viewers such as Mosaic. Hardware and software standards such as the instruction set architectures for microprocessor-based computers, modem protocols, media formats, and query languages also have played critical roles.

Standards can simplify many of the traditional data management jobs. For example, the time that would be used to decipher a tape format is saved and the job of installing a new application is facilitated. Having effective standards in place reduces the level of tedious, nonproductive effort and frees up time for new tasks for the archivist. Standards determined now will typically be in effect for long periods of time, perhaps a decade or more, with some small evolutionary augmentations. This means that a baseline of appropriate standards can be selected for a body of information with some reasonable expectation that they will not be quickly replaced. When it appears that the existing standards baseline needs to be updated, the information can then be migrated to a new one. A deliberate data migration strategy based on standards tracking is possible.

The role of standards certainly is not limited to the general computing community. Scientific teams and discipline groups continuously work to codify best practices, definitions, and algorithms. These are propagated as community standards. Standards developed by the scientific community are often the most important to promote and apply. If properly promulgated, they can enable improved understanding, broader collaboration, and facilitation of the data management and related research.

Finally, it should be emphasized that standards and guidelines to support long-term archiving must not inhibit innovation, or the evolution of information systems and technology. Often the best standards and guidelines are those that are independent of technology.

OPPORTUNITIES FOR NEW ORGANIZATIONAL STRUCTURES

With rapid technological improvements and newly enabled capabilities, it is sometimes easy to forget the importance of long-term commitment by managers to policy and resource requirements. No technological changes will by themselves replace the basic, unsung efforts of high-quality scientific data management. In fact, although technology itself can improve the availability of data, truly accessible and useful scientific information will be achieved only through such management commitment. This commitment must be based on a coherent strategy for life-cycle management of data, including technology acquisition, data and information management practices, and technology-independent standards to ensure that the minimum levels of data content and consistency for research uses are met. Further, such a comprehensive strategy will be successful only with the active and committed involvement of the scientific community itself. The level of effort and change that may be required to achieve this community involvement cannot be underestimated, and fundamental change to the value system of the community may be required.

Nevertheless, as discussed above, technological advances allow the creation of new infrastructure, challenging existing organizational assumptions. Effective organizational designs based on new allocations of responsibility are enabled. For scientific data management, the technological changes support organizations with the following attributes:

- *Widely distributed responsibility.* New telecommunications, data management, and standards technology allows for high levels of trust in distributed data management. Physical possession of data by archivists is no longer essential. The wide availability of information technology professionals and other skilled data managers (along with the lower technical skill levels actually needed) enhances the ability to distribute the data more broadly and increase user participation. Such distribution of data and their ownership (whether actual or implied) by user groups improves the utility of the data and helps create important support for long-term retention.

- *High-value peer-to-peer communication.* With access to data and to people on line, a variety of new collaborative relationships can develop. Information can be broadcast to interested individuals in a timely fashion. Data can be provided directly to field researchers to focus new data collection. Physical proximity and formal lines of communication are no longer vital to effective organizational operation. Indeed, closed, highly structured organizations often will be uncompetitive or fail to take full advantage of innovation.

- *Specialized data centers.* Distribution of resources implies that some specific locations can specialize and yet still contribute effectively to all. Specialized groups or institutions could be created in a scientific discipline or in some aspect of data management, archives, or standards. Designation of such specialized centers, in addition to those already in existence, is a significant mechanism for achieving economies of scale, reducing overall costs while enhancing the effectiveness of certain functions for the benefit of all.

- *Explicit long-term (technology) strategies.* A long-term technology strategy needs to be developed. The rapidly changing base of technology requires that a deliberate sequence of phases be selected, through which data and data management will migrate. The constant evolution of information technologies demands that an organizational element take on this "technology navigation" function.

- *Measurement as a vital tool.* In a fast-paced, and, perhaps, widely distributed effort, metrics are important to clearly communicate expectations of performance, register results, and help in detecting weak spots for corrective action. In particular, metrics could be established to determine data set use and to support archiving strategy decisions. Metrics also could be developed to help ensure high-quality service and proper data protection.

5
A New Strategy for Archiving the Nation's Scientific and Technical Data

The scientific and technical data held by federal government agencies and by other institutions supported by federal funds constitute an extremely valuable national resource. Unfortunately, in many cases this resource can be exploited only with great difficulty because key elements of the infrastructure for broad and easy access to it are incomplete or missing.

Currently, the most important development within the federal government for improving the management and long-term retention of scientific and technical data is the National Information Infrastructure (NII) initiative. The NII focuses on the application of public, private, and academic resources to define, implement, and maintain an evolving network of knowledge resources (IITF, 1993). This infrastructure will be the foundation for information-centered enterprises of the next century (NRC, 1994). The scientific community, whose lifeblood is widely available data and information, must become fully engaged in this national effort. A coherent strategy needs to be defined and implemented, to combine new technological capability with a new way of doing business throughout all phases of the scientific information life cycle (observation, measurement, analysis, interpretation, application, dissemination, and education).

An effective information infrastructure must build on enabling technologies to create an integrated and adaptive system that is easily accessible to all potential users. Each user community will have its own view of what the NII means to its enterprise and how the NII can best serve its users because the NII will be made up of many separate "enterprise information infrastructures." The existing scientific and technical data centers and archives already constitute a separate enterprise information infrastructure, which must become fully integrated into the NII.

In the discussion that follows, the committee lays out a three-part strategy for the long-term retention of scientific and technical data. The elements of this strategy are based on the technological advances outlined in Chapter 4 and on the issues raised in Chapter 2, which provide the context and the need for action.

The strategy begins with a set of fundamental principles for the long-term retention of scientific and technical data. The second major element outlines the committee's proposal to form a National Scientific Information Resource Federation, which would provide a coordination mechanism for end-to-end management of networked scientific and technical data facilities. The final sections highlight some specific recommendations for NARA and NOAA in their long-term retention of scientific and technical data.

FUNDAMENTAL PRINCIPLES FOR LONG-TERM DATA RETENTION

In order to respond adequately to the imperatives for preserving data about the physical universe and eventually to create an integrated, adaptive, and accessible infrastructure, the federal government should help establish effective and affordable processes for providing ready access to the vast national resource of scientific and technical data and related information. The process must support the needs of data originators, users, and custodians across all phases of the data life cycle, from origin to use by future generations. The committee believes that the following principles should guide the effort of the government agencies in the long-term retention of scientific and technical data:

- *Data are the lifeblood of science and the key to understanding this and other worlds. As such, data acquired in federal or federally funded endeavors, which meet established retention criteria, are a critical national resource and must be protected, preserved, and made accessible to all people for all time.* The original collection and analysis of scientific and technical data traditionally have been used primarily to support the scholarly publication of scientific interpretation by individual investigators. The availability of complete and consistent data sets for broader uses, both within and outside the scientific community, would significantly increase the return on the investment made in obtaining those data and provide insights not attainable if the original data were lost or unusable.

- *The value of scientific data lies in their use. Meaningful access to data, therefore, merits as much attention as acquisition and preservation.* Technology can make data available through fast computers, large-bandwidth networks, massive storage capabilities, and portable media. However, if the paths to data are obscure, or there is no way for a user to determine what is significant and relevant, then the data become inaccessible and are effectively lost.

- *Adequate explanatory documentation, or metadata, can eliminate one of today's greatest barriers to use of scientific data.* The problem of inadequate metadata is amplified when users are removed from the point of origin by being in a different discipline, by having a different level of expertise, or by time. Addressing this problem comprehensively will make data useful in the broadest possible context.

- *A successful archive is affordable, durable, extensible, evolvable, and readily accessible.* These terms may appear to be vague targets, but they imply basic goals. The costs of developing, operating, and using an archive must not be excessive. The archive must endure the ravages of long-term use, and it must be able to extend broadly the services it offers and the records it manages. It must evolve to support the assimilation of new technology, policies, procedures, and uses. Finally, an archive is not effective if a broad population of users cannot use it. The archiving system thus should provide multiple levels of access to any subset of its holdings, although holdings not accessed often may not require a sophisticated access mechanism.

- *The only effective and affordable archiving strategy is based on distributed archives managed by those most knowledgeable about the data.* Archive centers generally should be at the agencies or institutions that collect the data, and they should be responsible for archiving and providing access to the data as long as the agency's or institution's mission and scientific competence continue to encompass the subject field. Physical transfers of the data should be avoided if possible, so agencies and institutions will need to allocate adequate resources to the entire life cycle of their data holdings.

- *Planning activities at the point of data origin must include long-term data management and archiving.* This principle is recognized in the Office of Management and Budget Circular A-130 on the "Management of Federal Information Resources" (OMB, 1994). The scientific information management spectrum spans data collected from a sensor to the scholarly publications that report scientists' interpretations of the data. Scientists, information technology professionals, data managers, librarians, and archivists must unify their expertise in the establishment of a coherent strategy for end-to-end data and information management. Although these communities traditionally have not worked closely together, their combined knowledge and effort are now required. The benefit of incorporating planning at the point of origin is that it is cheaper and more effective to plan for retention than to reconstruct data sets later.

THE PROPOSED NATIONAL SCIENTIFIC INFORMATION RESOURCE FEDERATION

The committee believes that the federal government should create a National Scientific Information Resource Federation—an evolutionary and collaborative network of scientific and technical data centers and archives—to take on the challenge of providing effective access to and preservation of important scientific and technical data and related information. Such an initiative would begin to exploit more fully our nation's significant investment in the physical (and other) sciences and the data acquired with that investment. In the discussion that follows, the committee reviews the basic elements of a federated management structure, describes some notable examples of existing federal government organizations for large-scale distributed data management, and outlines the most important aspects of the proposed National Scientific Information Resource Federation.

Elements of a Federated Management Structure

Several critical concepts must govern any federated management structure for it to function properly. These include the notions of subsidiarity, pluralism, standardization, the separation of powers, and strong leadership at all levels (Handy, 1992).

Subsidiarity means that power is assumed to lie with the subordinate units of an organization and can be relinquished, but not taken away. The subordinate units typically are best qualified to make operational decisions that directly affect them and that they will be implementing. The central management is allowed only those powers needed to ensure that the subordinates do not damage the organization. For example, the Constitution of the United States reserves only specified powers for the federal government, with any unstated powers belonging to the states. Applied to the situation at hand, it is clear that the strengths of the current system for managing scientific and technical data and information in the United States are distributed among a number of diverse data centers and archives, both within and outside the government. A successful federation of these existing institutions would recognize that they are the locations of expertise on their respective data holdings. Thus the central organization should be small and should not micromanage the day-to-day operations of the subsidiary organizations.

Pluralism may be defined as interdependence of the members. In a federation, the individual subsidiary organizations recognize the advantages of belonging to the federation, because of products or services that can be obtained from other elements in the federation. As noted in the previous chapter, the existence of many specialized data centers and archives, as well as the possibility of creating new ones in a networked environment, can offer significant economies of scale and improved sharing of ideas and expertise. What is good for the subsidiary element also should be good for the whole. Pluralism, coupled with subsidiarity, guarantees a measure of democracy in the federation.

Interdependence, in turn, requires **standardization** of languages, communications, basic rules of conduct, and units of measurement. These elements may be summarized as technical and procedural standardization. This too was discussed in Chapter 4, regarding the development of standards in software, hardware, and data management. Standards that are developed by consensus of the subsidiary elements (e.g., the participating data centers, archives, and researchers) are widely recognized as essential to the successful management of data.

A separation of powers (responsibilities), with a system of checks and balances, is necessary to ensure that the central authority does not take on unnecessary power. This principle must be incorporated into the federation's organizational structure.

Finally, a federation requires **strong leadership** that is effective, yet not overbearing. The central coordinating element or executive office must act as the standard bearer, promoting the federation's established goals and objectives while reminding the subsidiary organizations of the importance of carrying out their responsibilities.

Examples of Distributed Data Management Organizations

Successful examples of a federated management structure are numerous in the private sector (Handy, 1992). More specifically, however, there already are two large-scale, federal government, distributed data management groups that embody many, though not all, of the federated management attributes outlined above. These are the Interagency Working Group on Data Management for Global Change and the Federal Geographic Data Committee.

Interagency Working Group on Data Management for Global Change

In 1990, Congress formally established the U.S. Global Change Research Program (GCRP), "aimed at understanding and responding to global change, including the cumulative effects of human activities and natural processes on the environment, [and] to promote discussions toward international protocols in global change research . . ." (CENR, 1994). The activities of the GCRP are coordinated by the Committee on Environment and Natural Resources (CENR), under the President's National Science and Technology Council.

The timely availability of a broad spectrum of scientific data and information, from both governmental and nongovernmental sources, is fundamental to meeting the goals of this program. A Global Change Data and Information System (GCDIS) is being created to facilitate access to and use of the data and information necessary to support global change research. The federal organizations involved in the GCDIS planning include the Departments of Agriculture, Commerce, Defense, Energy, Interior, and State, as well as the Environmental Protection Agency, the National Aeronautics and Space Administration, and the National Science Foundation.

According to *The U.S. Global Change Data and Information System Draft Implementation Plan* (CENR, in press), the GCDIS is building on the resources and responsibilities of each participating agency, linking the data and information services of the agencies to each other and to the users. The system thus is composed largely of the separately funded components contributed by the participating agencies. It is supplemented by a minimal amount of crosscutting new infrastructure through the use of standards, common management approaches, technology sharing, and data policy coordination. Neither a lead agency nor a separately funded budget for the GCDIS is planned; rather, implementation of the system is being coordinated through the Interagency Working Group on Data Management for Global Change (IWGDMGC). Decision making, therefore, is done through a consensus process based on the common interests of all participants.

Plans for the GCDIS recognize that the global change data must be available for a very long time, regardless of the changing interests of the researcher, group, or agency that originally collected and analyzed the observations. Although each agency participating in the GCDIS is expected to manage, store, and maintain the data sets under its purview, the plan does allow an agency to designate another GCDIS agency to archive some of its data. The participating agencies are expected to adhere to government standards for media, storage, and handling as prescribed by NARA and the National Institute of Standards and Technology. The agency archives associated with the GCDIS access system will be staffed by professionals who understand the data and their sources. The IWGDMGC expects to develop guidelines for preparing data sets and associated documentation for long-term retention at the participating agencies. Ideally, the GCDIS archives also will be associated with research groups, both within and outside government, who, as principal users of those data, will verify quality and documentation of the data.

The GCDIS plan gives each agency responsibility for its own data-purging policies, although interagency coordination procedures will be developed to prevent the loss of important data sets. Before any data sets are purged, however, an agency will be required to notify the IWGDMGC of its plans at least one year in advance, and to allow other GCDIS agencies to indicate their requirements for those data, or to agree to assume responsibility for the archiving of those data. In the event that no agreement can be reached on the disposition of a data set identified for purging, existing NARA procedures will apply (CENR, in press).

Federal Geographic Data Committee

The other major federal data coordination entity important to the long-term management of observational data (including some data from the biological and social sciences) is the Federal Geographic Data Committee (FGDC). The Office of Management and Budget (OMB) established the FGDC in 1990 to develop a National Spatial Data Infrastructure (NSDI) to work toward the coordinated development, use, sharing, and dissemination of geographic data (OMB, 1990). Participating government organizations include the Departments of Agriculture, Commerce, Defense, Energy, Housing and Urban Development, Interior, State, and Transportation, as well as the Environmental Protection Agency, Federal Emergency Management Agency, Library of Congress, National Aeronautics and Space Administration, National Archives and Records Administration, and Tennessee Valley Authority. In fulfilling its mandate, the FGDC carries out the following activities, among others:

- promotes the development, maintenance, and management of distributed database systems that are national in scope for geographic data;
- encourages the development and implementation of standards, exchange formats, specifications, procedures, and guidelines;
- promotes technology development, transfer, and exchange; and
- promotes interaction with other existing federal coordinating mechanisms that have interest in the generation, collection, use, and transfer of spatial data (FGDC, 1994).

The FGDC has received authority and some limited funding to pursue these objectives. Specifically, Executive Order 12906 on "Coordinating Geographic Data Acquisition and Access: The National Spatial Data Infrastructure," assigns to the FGDC the responsibility to coordinate the federal government's development of the NSDI. That Executive Order also instructs the FGDC to involve state and local governments in its NSDI activities, and to use the expertise of academia, professional societies, the private sector, and others as necessary to assist the FGDC.

The FGDC has established a matrix of subcommittees and working groups according to discipline-related data categories and interests. The working group issues include a framework for data, a clearinghouse for data, standards, technology, and data archiving. The FGDC plans for data archiving are still being developed, however.

Creation of the National Scientific Information Resource Federation

The two examples cited above indicate that a federated management structure for highly distributed scientific data can be created. In fact, between these two groups, the life-cycle management of many of the data that are the topic of this report is beginning to be systematically approached. Nevertheless, as discussed in this report and in the volume of working papers (NRC, 1995), many important gaps and inadequacies remain in the management and retention of our nation's scientific data and related information. The committee believes that these deficiencies can best be addressed by a comprehensive federated system—a National Scientific Information Resource (NSIR) Federation—that builds on the successes of the existing groups and helps coordinate them with other data management entities that still need improvement.

There are many reasons why it is now propitious to establish a system of federated data management, with an emphasis on long-term retention. From a policy perspective, it would be consistent with the goal of the National Information Infrastructure to distribute information resources broadly throughout our society, with the federal government acting as facilitator for such activities. The technology is available to make a fully networked, but highly distributed, system of data centers and archives both feasible and desirable. Such a system would be efficient in providing access to scientific data and information to a large number of potential users and would maximize the government's return on the significant investment that initially went into acquiring those data. From an organizational standpoint, a federated management structure would allow the disparate elements to continue to specialize in what they each do best and to fulfill their individual organizational mandates, while providing some efficiencies of scale and political leverage in addressing the most pressing issues. Moreover, this type of approach is especially timely and important in an era of federal government budget reductions. The committee therefore envisions a broadly networked organization, which would be implemented through the collaboration of the federal government's scientific and technical agencies as well as commercial and noncommercial organizations outside the government, and integrated into the emerging National Information Infrastructure.

Most of the elements of the NSIR Federation are already in place. These include the data centers and field archives run by several of the federal agencies that are among the primary generators and collectors of the nation's scientific data and information. In addition to holding data, these centers and archives have highly skilled staff with the requisite expertise. The organizations are widely distributed, both geographically and by discipline.

The existing data centers and field archives, however, do not approach the federated organizational model for several reasons. There is no unifying organization among the various elements, there is wide disparity in the quality and depth of service provided, and few of them have a charter to preserve data "permanently." Although NARA has the statutory charter to preserve federal records in perpetuity, its current and projected holdings of electronic scientific records are very small. While the committee does not believe that NARA's archives of scientific data should increase substantially, it found little evidence of activity within the scientific and technical agencies that would indicate that their ability to provide for long-term retention and access to their data would improve without some restructuring.

A fundamental precept is that those most familiar with scientific data—the scientists themselves—are in the best position to oversee the management of those data (NRC, 1982). In light of the volume and diversity of scientific data, a distributed approach that maintains the data closest to the primary user community is the most effective method for managing them. As mentioned above, several agencies have adopted an approach of caring for their data in systems of field archives or discipline data centers. Although these agencies have devoted significant attention to the preservation of data, their concern is limited to providing immediate service to primary users of the data for their originally intended purpose. Little thought has been given to the perpetual archiving of the data within most agencies, with the notable exception of NARA and NOAA, which already have a statutory mandate that allows them to preserve data collected by the federal government. Because it is not possible to be sure that any data center will exist in perpetuity, some mechanism must be in place to ensure that the data will be retained by an appropriate organization—within or outside the government—in the event that the continued existence of a data center is jeopardized.

If a lead agency can be determined for a subject matter, then it should take responsibility for coordination of scientific data on that subject, no matter which agency has physical ownership or custody of those data. The committee recognizes, however, that some data sets are largely of interest at the boundaries of disciplines or agency charters and that consequently these may be more difficult to manage or document properly. Large data sets that are of an interdisciplinary nature cause special problems in this regard. For these complex situations, no simple rule will take the place of negotiations among the involved agencies to make the necessary arrangements for long-term archiving. Indeed, every agency should assume the obligation to keep its holdings of scientific data in usable form, even if the data are not in active use, until agreeing on disposition of those data with NARA or another agency.

In addition to the agency-administered data centers, there are educational or private concerns that hold and administer data important to one or more agencies, such as the archived data from the NOAA Geostationary Operational Environmental Satellites at the University of Wisconsin or the seismic data held by the Incorporated Research Institutions for Seismology. While some of these nonfederal archives are firmly associated with one or more federal agencies through contractual and funding relationships, in other cases a one-to-one association is less clear. It follows that a well-defined chain of responsibility must be established for all data that are to be preserved. This decision should be made by the individuals and institutions most closely associated with and interested in those data, and it should be made with due consideration for cost efficiency, appropriate expertise, scientific interest, and convenience, among other factors. Establishing a clear connection between a field archive and an agency should in no way limit the community of users served by the archive, but should ensure an orderly and secure path of responsibility for the data.

The structure of the nation's scientific and technical organizations continues to change. In some instances, institutions or even agencies will merge, while in other cases, organizations may disappear. When such changes occur, it is likely that the scientific interests formerly represented by those organizations will be subsumed by existing or new agencies or organizations. The general topology of the NSIR Federation, however, would not change.

The committee does not anticipate that the creation and implementation of the Federation will require much additional funding, if any, because it will consist primarily of improving linkages and coordination among existing data centers, archives, and related organizations within a highly decentralized management structure. Moreover, any costs incurred in this process should be more than offset by the improvements in efficiency and access to the data and related information resources.

RECOMMENDATIONS FOR THE CREATION OF THE NSIR FEDERATION

The committee thus recommends that the federal government take the following steps for adequately preserving and providing access to data about our physical universe:

Adopt the National Scientific Information Resource (NSIR) Federation concept as an integral part of the National Information Infrastructure (NII). This concept must encompass not only an electronic network, but also individuals, organizations, communities, data resources, procedures, guidelines, and associated activities of data generation, management, custodianship, and use. The NSIR Federation should provide the foundation for defining a coherent approach to management of the life cycle of scientific data, with the goal of providing broad and effective access to all potential users as cost effectively as possible. The Federation should be developed and implemented through consensus of collaborating organizations with diverse and autonomous missions. The GCDIS, in particular, is an example of a prototype NSIR, focused on data for a specific set of interdisciplinary science problems. The NSIR Federation would build on such efforts, providing for better coordination and interaction among them, and would help organize fledgling efforts to preserve and provide access to data in other disciplines.

The administration should take the steps necessary to fully define and create the NSIR Federation. There are at least two potential focal points within the administration for planning such an activity. These are the interagency Information Infrastructure Task Force for the NII and the National Science and Technology Council. The NSIR Federation could be created in a manner similar to the creation of the Federal Geographic Data Committee and its National Spatial Data Infrastructure (e.g., through an Office of Management and Budget Circular and Executive Order), or of the Interagency Working Group on Data Management for Global Change and its Global Change Data and Information System (e.g., through legislation in cooperation with the administration). A convocation of representatives from the scientific, data and information management, and archiving communities would be a good way to define and inaugurate this initiative, focusing on the most significant issues and problems identified at the end of Chapter 2.

Following the formal authorization by the federal government for creating the NSIR Federation, the principal parties, including NARA and NOAA, should conclude agreements for the implementation of a distributed archive system. The system should involve all relevant institutions, including nongovernmental entities that are funded by the federal government or that maintain data that were acquired with federal funds. As a general principle, data collected by an agency should remain with that agency indefinitely. The committee recognizes that this recommendation may require significant operational changes for agencies other than NOAA, and even some changes with respect to NOAA's data activities. In addition, NARA should consider concluding interagency agreements to give formal recognition of this process as appropriate. Furthermore, the associated agencies in the NSIR Federation must work together, under the lead of a small, coordinating executive office with the expertise to establish data management guidelines and minimum criteria for adequate metadata that could be applied across the entire Federation. The executive office could be either a high-level interagency coordinating committee, similar to the FGDC, or a new office at an appropriate federal agency, such as the National Science Foundation, which has a broad scientific and technical as well as communication mandate. In any case, the executive office should resist the typical tendency toward bureaucratic accretion of power, personnel, and resources, and the tendency to consolidate and centralize data holdings. A management council consisting of representatives of the member organizations should be created to help ensure that the central executive function remains fully responsive to all members of the Federation.

Data access and preservation services should be implemented on the most cost-effective basis possible for the Federation. For example, one institution may provide a service to one or more other institutions in order to exploit potential economies of scale and focal points of expertise (e.g., the specialized data centers suggested in Chapter 4). This measure might increase the cost to the providing institution, but would decrease the overall cost to the federation, the government, and the taxpayer. An example of this is the method by which backup copies of data might be kept. NARA may have at any given time the most cost-effective "vault" in which to keep physically separate backup copies of data for all agencies, and, hence, the federal government would save money by increasing NARA's budget to provide this service for the other agencies. On the other hand, if cost trade-off studies were to find that a single large "vault" is not as cost-effective as distributed facilities, then each agency would be responsible for its own backup. In all NSIR Federation activities, emphasis should be placed on control of costs, with the most successful methods used by individual members identified and shared with all other members.

The institutions belonging to the NSIR Federation should develop a process for collaborating effectively on specific initiatives. This process should provide a mechanism to define and prioritize data management and preservation initiatives, to establish the required agreements between collaborating organizations, and to secure funding for each initiative. Each participating organization would contribute to the Federation according to its particular strengths and in a manner consistent with the founding charter. In addition, an independent advisory body consisting of experts from user groups should be formed in support of each initiative.

The NSIR Federation should develop a national resource of information technology that is consistent with its chartered objectives and that can be effectively distributed to institutions that must manage data. These technologies would include complete products, designs, guidelines, standards, and methodologies. A related long-term technology strategy, or "technology navigation" function, should be developed, as suggested in Chapter 4.

The NSIR Federation should institute an independently managed process for awarding NSIR certification to member scientific institutions and their data and information systems on the basis of well-defined criteria and standards. The certification process should be managed by a nongovernmental, not-for-profit organization, which would receive technical guidance from the participating federal agencies. The certification needs to have credibility in the community so that nonmember institutions will aspire to attain certification and have it tagged to their products. The certification also should be something that commercial value-added providers will seek to increase the credibility of their products.

It also is important for the committee to state what the NSIR Federation should not be. It should not become an expensive bureaucratic entity. The executive office must not impose any standards or information technologies from above that have not been validated through a consensus process of the member organizations. Finally, the executive office must not attempt to micromanage the operations of the participants, nor should it have any direct control over their budgets and funding allocations.

RECOMMENDATIONS SPECIFICALLY FOR NARA

In order to improve its responsibilities in the long-term retention of scientific and technical data, the committee recommends that NARA strengthen its liaison with each federal agency that produces such data to ensure that appropriate attention is devoted to long-term data retention in a distributed storage environment.

As shown earlier in this report, NARA cannot today, nor will it likely ever be able to, act as the custodian of most physical science data. The data volume is too great in relation to the funding appropriated to NARA, the NARA staff do not have the necessary specialized scientific knowledge, the interagency linkages are not in place, and a huge infrastructure similar to that which already exists at other agencies would need to be duplicated at NARA. The agencies closest to the data sets and best equipped to deal with them are themselves already struggling with these issues. However, NARA does have great expertise in issues involving the long-term storage of data and the packaging requirements for data to be of value to future users.

The committee therefore believes that NARA's role should be primarily advisory or consultative, to help ensure that the agencies that are the actual custodians of data at the working level follow all the relevant federal laws and guidelines in taking care of the data. The committee suggests that scientific data and related information should go to NARA's physical possession only as a last resort, when the agency that collected the data can no longer provide access for the user community. As has already been noted, scientific data are best maintained by the agency that originally acquired those data as long as there is any regular active use. The holding agencies should collect, analyze, store, and make available the maximum feasible amount of relevant physical science data, consistent with the principles and goals set forth for the NSIR Federation and with the retention criteria and appraisal guidelines discussed above.

Currently, agencies inform NARA of their intentions for their federal records, including scientific data, through various schedules. All agencies are required to schedule records when they reach 30 years of age, although they are encouraged to do so earlier. The National Climatic Data Center even provides schedules for data that it plans to hold indefinitely, noting that intention. For most types of records, the pressure to schedule provides the useful function of preventing an agency from simply warehousing continually increasing volumes of unused records without examination. For data that an agency does not wish to destroy, but that are not frequently accessed, NARA makes available storage space without taking ownership. If NARA did not provide some worthiness test for records before agreeing to provide storage for another agency, the Federal Records Centers could become inundated with records of little value or potential for future use.

As discussed in this report, we are heading increasingly toward a system of distributed archives for electronic records. Data sets are distributed among various physical locations, and the expertise to interpret these data sets is likewise already distributed and becoming more so. The rapid increase in computer networks within the United States and in the rest of the world is beginning to significantly affect the way people access information. There is a lessening need for data users and providers to physically possess the data they need or distribute, and users are increasingly unaware of the source location(s) of the data they are accessing. NARA therefore should continue to study arrangements regarding the physical custody of electronic records, the relationship between NARA and other agencies, and how these will and should be affected by the expansion of electronic networks.

During the course of this study, the committee found that with the exception of some staff members at government data centers, many government scientists and most nongovernment scientists are not aware of the requirements of the Records Disposal Act (44 U.S.C. 3301 et seq.). Even some of those entrusted with large quantities of valuable data were largely unaware of NARA and its related responsibilities until contacted by the committee, or by its panels. This may be partially because scientists, even those within the federal government, sometimes do not respond to the bureaucratic requirements of their own institutions. The committee is encouraged that NARA is working to address this problem. Nevertheless, many panel visitors and members observed that the NARA brochures have an authoritarian and legalistic tone and are not conducive to establishing productive partnerships with NARA. NARA's future effectiveness in overseeing and advising on the archiving of scientific and technical data requires that it improve its relations with other agencies and institutions.

As a corollary, none of the committee's suggestions should be construed to imply that NARA should issue additional proclamations or regulations. The goal should be to present more carrots than sticks. For example, NARA should consider providing rewards and recognition to researchers, managers, and funders for developing and implementing successful data retention plans, with appropriate metadata. With better communications and greater sensitivity to the needs of the scientific community, NARA can play the role of a "service provider" and "appraisal consultant." For instance, NARA is already working with the DOD Legacy Resource Management Program to identify and preserve cultural resources under DOD jurisdiction. NARA and this DOD program together have sponsored a conference to assist military contractors in preserving their documentary heritage. The committee suggests that NARA pursue other such collaborations in the same spirit of partnership.

As a matter of formal responsibility and training, NARA staff are more concerned with long-term archiving issues than most staff at other agencies. NARA therefore can serve an essential role in reminding agencies of the long-term value of data and should regularly provide advice to agencies that keep scientific data on hand for extended periods of time. NARA also should conduct continuous research on retention and appraisal issues to remain well-informed. **The committee recommends that NARA form standing advisory committees with managers of scientific data, historians, and scientific researchers to address the retention and appraisal of scientific and technical data collections, and related issues.**

Unfortunately, NARA has almost no scientific expertise within its ranks (except related to physical records preservation). Despite the large amounts of scientific information within some federal records, NARA officials have indicated that they do not believe that they could keep a scientist on the staff interested in the work and do not plan to hire any permanent scientific personnel. Nevertheless, NARA will continue to be faced with difficult issues involving the archiving of scientific data. In the interim, the committee suggests that NARA should arrange for temporary staff assignments from the active scientific ranks of the federal government on a frequent as-needed basis. Given the great challenges that NARA will face from scientific data and the proven ability of other agencies to hold scientifically trained personnel in data management positions, NARA should rethink its position and consider creating a cadre of permanent staff with scientific expertise.

NARA also might consider setting up an in-house database to track federal holdings, especially to anticipate problems with data sets housed in other agencies that may eventually need NARA protection or other help from NARA. To do this effectively would require establishing a set of contacts in other agencies with people who understand the databases in the agency collections.

This brings us to the need for a more general locator function, or "directory of directories," for the NSIR Federation's network of networks. Archives must not be viewed or managed as data cemeteries, with only rare and dwindling visits after the deposition of data. The provision of broad access to data must be part of archive design and construction, and thus some sort of broad locator is much needed. The committee is encouraged by the recent interagency efforts, organized by the Office of Management and Budget, to develop a Government Information Locator Service. Nevertheless, there is a need for a NARA-maintained directory of archived data within its own system. This should include archived records maintained by other government agencies and federally funded institutions that are recognized as part of a distributed archive system overseen broadly by NARA. **The committee recommends that NARA collaborate with other agencies that maintain long-term custody of data to develop an effective access mechanism to these distributed archives. The initial step should focus on locator systems and evolve toward a transparent access system.**

Finally, with regard to its requirements for accession of data, NARA should work with the scientific community and potential sources of scientific data to develop adaptable *performance* criteria for data formats and media, rather than mandating narrow and inflexible product standards. The goal would be to meet NARA's basic need to ensure long-term usability while also enabling accession of data, such as images and structures, that cannot be accommodated by NARA's current restrictive file-format and media standards.

RECOMMENDATIONS SPECIFICALLY FOR NOAA

As the largest holder of earth sciences data in the United States, NOAA has a vast amount of scientific data stored at many facilities across the country. The primary storage sites are the National Data Centers, which include the National Climatic Data Center (NCDC), the National Oceanographic Data Center (NODC), and the National Geophysical Data Center (NGDC). Each of these data centers now has its own on-line information service. The data centers are accessible through common nodes, for example through NOAA's web server or NASA's Master Directory server. Thus a user who understands the structure of NOAA's data holdings can navigate through the different data centers, look for data of interest in each center's holdings, and retrieve the data over the Internet. However, it is not possible to search NOAA's data holdings with the same precision and accuracy with which one can search for bibliographic data, through, for example, the Current Contents or INSPEC databases. The diversity and volume of data that the National Data Centers hold and regularly receive make it difficult to produce an overall directory for all of NOAA's data holdings. In particular, NCDC receives daily all of the weather information for the United States. Without such a general directory it is difficult for users to query across NOAA archives to locate and integrate diverse data. Moreover, once the user finds data, the variety of storage formats and data types makes access cumbersome. Thus, the committee encourages NOAA to be ambitious. Development of a new comprehensive directory covering all NOAA's holdings of geoscience data would set the standard for other agencies and would make the data much more accessible to the public.

This directory may incorporate capabilities of the many different on-line directory services currently in use at the National Data Centers, but the emphasis should be on connectivity, data access, and information. For this reason, NOAA should concentrate first on the more recent digital data that can most easily be incorporated into such a directory system. Efforts to get older analog data digitized should continue, although some data may have to remain in their original format. An important facet of this directory is to list, along with the directory entry, how to locate and access the data. Once they have located the data of interest, most users want mainly to retrieve the data in a form that they can use for further analysis.

Thus, the directory should specify the actual location of the data, as well as the methods by which the data can be acquired. Under the present NOAA system, acquisition involves a formal ordering procedure and the transfer of funds, at least for any data that must be transferred via tape or hard copy. Experimental NOAA systems (NOAA's Satellite Active Archive) make it possible to order limited satellite imagery over the network at no cost. For those orders requiring the transfer of funds, the directory service should be able to estimate the cost of the data order so that the user can factor cost into the decision to order.

This interconnected NOAA directory service also would assist the NOAA data centers in their management of data. By having access to tools and techniques developed at other NOAA data centers and elsewhere in the data storage community, the NOAA data centers would be better able to stay abreast of new developments and to incorporate them into their data access systems. Similarities among various earth science data and the emerging need for interdisciplinary research make it necessary to implement such an overall directory for managing NOAA data, for both data location and access. As noted earlier, NOAA already has started to develop data directories, on-line data systems, and data access.

NOAA and NASA have made progress in data rescue and in deriving better products from old data. Since 1990, NCDC has copied thousands of tapes of satellite data that were at the end of their useful shelf life. The NOAA/NASA Pathfinder program was established to make the satellite data more generally available to researchers and to calculate new products; it has been an effective program. Although the committee supports activities to preserve old data, rescued data (including data moved to better media and analog data that have been digitized) are of little value if they cannot be accessed or retrieved. The committee advocates more emphasis on improving access to data for interested users.

Most federal agencies are now aware that storage and retrieval of data are important. Problems arise because each agency, and sometimes even different parts of the same agency, sets up data centers and facilities, and each of these establishes its own type of system. In addition, because the technology for storing data changes frequently, it is difficult if not impossible to decide just what hardware and software system should be used. This uniqueness of systems often hinders system portability and the exchange of data among systems.

There are some approaches and procedures that are designed to be technology-independent and therefore can be used to avoid some of these problems. Moreover, the technological and portability requirements for archiving, storage, and transmission are different, so a "universal" format will not work. An **archival format** must be utterly portable and self-describing, on the assumption that, apart from the transcription device, neither the software nor the hardware that wrote the data will be available when the data are read. A **storage format** should be optimized for retrieving any addressable subset of a dataset. A secondary, but important, consideration is the ease with which the storage format may be cast into a **transmission format**. A transmission format should be optimized for ease of conversion to other formats, accommodation of both data and metadata in a single data stream, portability, and extensibility (i.e., accommodating data and metadata types and structures not yet invented). Because both NOAA and NARA have a long-term archival problem, the committee suggests that they work together to locate and test hardware and software units that can be used for this technology-independent approach. By locating the most simple common technologies, it should be possible to set up systems that are sufficiently capable, but yet are able to interact with each other. Once a few of these "standards" are set up and operating, it is likely that other users will want to run this suite of software. Ideally, this type of project would be best carried out under the auspices of the NSIR Federation.

Considering the foregoing discussion, the committee makes the following recommendations:

NOAA should place a higher priority on documenting and establishing directories of its data holdings.

Furthermore, NOAA, with the active cooperation of NARA, should lead efforts to better define technology-independent standards for archiving, storing, and transmitting the data within its purview.

Finally, NOAA, as well as every other federal science agency, should ensure that all its data are shared and readily available; it fulfills its responsibility for quality control, metadata structures, documentation, and creation of data products; it participates in electronic networks that enable access, sharing, and transfer of data; and it expressly incorporates the long-term view in planning and carrying out its data management responsibilities.

The creation of the committee's proposed NSIR Federation would help provide a collaborative mechanism and more sustained peer pressure to meet these objectives, and thus enhance the value of scientific and technical data and information resources to the nation.

References

American Chemical Society (ACS). 1993. *Reporting Experimental Data*, H.J. White (ed.), Washington, D.C.

Boorstin, D.J. 1992. *The Creators*, Random House, New York.

Committee on Environment and Natural Resources (CENR). 1994. *Our Changing Planet: The FY 1995 U.S. Global Change Research Program*, National Science and Technology Council, Washington, D.C.

Committee on Environment and Natural Resources (CENR). In press. *The U.S. Global Change Data and Information System Draft Implementation Plan*, National Science and Technology Council, Washington, D.C.

Federal Geographic Data Committee (FGDC). 1994. October 1994 Fact Sheet, Federal Geographic Data Committee, Washington, D.C.

Gelsinger, P.P., P.A. Gargini, G.H. Parker, and A.Y.C. Yu. 1989. Microprocessors circa 2000, *IEEE Spectrum*, October: 43-47.

General Accounting Office (GAO). 1990a. *Environmental Data—Major Effort Is Needed to Improve NOAA's Data Management and Archiving*, Washington, D.C.

General Accounting Office (GAO). 1990b. *Space Operations—NASA Is Not Archiving All Potentially Valuable Data*, Washington, D.C.

Haas, J.K., H.W. Samuels, and B.T. Simmons. 1985. *Appraising the Records of Modern Science and Technology: A Guide*, Massachusetts Institute of Technology, Cambridge, Mass.

Handy, C. 1992. Balancing Corporate Power: A New Federalist Paper, *Harvard Business Review* 70(6): 59-72.

Information Infrastructure Task Force (IITF). 1993. *The National Information Infrastructure: Agenda for Action*, Washington, D.C.

Jacobs, W. 1947. Wartime developments in applied climatology, *Meteorological Monographs* 1(1), 52 pp.

Marshack, A. 1985. *Hierarchical Evolution of the Human Capacity: The Paleolithic Evidence*, American Museum of Natural History, New York.

National Academy of Public Administration (NAPA). 1991. *The Archives of the Future: Archival Strategies for the Treatment of Electronic Databases*, A report for the National Archives and Records Administration, Washington, D.C.

National Aeronautics and Space Administration. 1992. *Draft Guidelines for Development of a Project Data Management Plan (PDMP)*, NASA Office of Space Science and Applications, Washington, D.C.

National Research Council (NRC). 1982. *Data Management and Computation—Volume I: Issues and Recommendations*, Space Science Board, National Academy Press, Washington, D.C.

National Research Council (NRC). 1984. *Solar-Terrestrial Data Access, Distribution, and Archiving*, Space Science Board and Board on Atmospheric Sciences and Climate, National Academy Press, Washington, D.C.

National Research Council (NRC). 1986a. *Atmospheric Climate Data: Problems and Promises*, Board on Atmospheric Sciences and Climate, National Academy Press, Washington, D.C.

References

National Research Council (NRC). 1986b. *Issues and Recommendations Associated with Distributed Computation and Data Management Systems for the Space Sciences*, Space Science Board, National Academy Press, Washington, D.C.

National Research Council (NRC). 1988a. *Geophysical Data: Policy Issues*, Committee on Geophysical Data, National Academy Press, Washington, D.C.

National Research Council (NRC). 1988b. *Selected Issues in Space Science Data Management and Computation*, Space Science Board, National Academy Press, Washington, D.C.

National Research Council (NRC). 1990. *Spatial Data Needs: The Future of the National Mapping Program*, Board on Earth Sciences and Resources, National Academy Press, Washington, D.C.

National Research Council (NRC). 1992a. *Setting Priorities for Space Research: Opportunities and Imperatives*, Space Studies Board, National Academy Press, Washington, D.C.

National Research Council (NRC). 1992b. *Toward a Coordinated Spatial Data Infrastructure for the Nation*, Board on Earth Sciences and Resources, National Academy Press, Washington, D.C.

National Research Council (NRC). 1993. *1992 Review of the World Data Center-A for Rockets and Satellites, National Space Science Data Center*, Board on Earth Sciences and Resources, National Academy Press, Washington, D.C.

National Research Council (NRC). 1994. *Realizing the Information Future—The Internet and Beyond*, NRENAISSANCE Committee, Computer Science and Telecommunications Board, National Academy Press, Washington, D.C.

National Research Council (NRC). 1995. *Study on the Long-term Retention of Selected Scientific and Technical Records of the Federal Government: Working Papers,* Commission on Physical Sciences, Mathematics, and Applications, National Academy Press, Washington, D.C.

National Research Council (NRC). In press. *Finding the Forest in the Trees: The Challenge of Combining Diverse Environmental Data*, U.S. National Committee for CODATA, National Academy Press, Washington, D.C.

Office of Management and Budget (OMB). 1990. Coordination of Surveying, Mapping, and Related Data Activities, Circular No. A-16, Washington, D.C.

Office of Management and Budget (OMB). 1994. Management of Federal Information Resources, Circular No. A-130 (59 F.R. 37906, July 25, 1994), Washington, D.C.

Office of Technology Assessment (OTA). 1994. *Remotely Sensed Data: Technology, Management, and Markets*, OTA-ISS-604, Government Printing Office, Washington, D.C.

Silberschatz, A., M. Stonebreaker, and J. Ullman. 1991. Database systems: Achievements and opportunities, *Communications of the ACM* 34(10): 110-120.

Appendix A
List of Acronyms

CD-ROM	Compact Disk-Read Only Memory
CENR	Committee on Environment and Natural Resources
DMC	Data Management Center
DOD	Department of Defense
DOE	Department of Energy
EROS	Earth Resources Observing System
ESDM	Earth Science Data Management
FGDC	Federal Geographic Data Committee
FITS	Flexible Image Transport System
GARP	Global Atmospheric Research Program
GCDIS	Global Change Data and Information System
GCRP	Global Change Research Program
GILS	Government Information Locator Service
HTML	HyperText Markup Language
IRIS	Incorporated Research Institutions for Seismology
IWGDMGC	Interagency Working Group on Data Management for Global Change
JANAF	Joint Army-Navy-Air Force
JCL	Joint Control Language
NARA	National Archives and Records Administration
NCDC	National Climatic Data Center
NGDC	National Geophysical Data Center
NII	National Information Infrastructure
NOAA	National Oceanic and Atmospheric Administration
NODC	National Oceanographic Data Center
NRC	National Research Council
NSDI	National Spatial Data Infrastructure
NSF	National Science Foundation
NSIR	National Scientific Information Resource
NSSDC	National Space Science Data Center
OMB	Office of Management and Budget

Appendix A

PDS	Planetary Data System
PO.DAAC	Physical Oceanography Distributed Active Archive Center
SGML	Standard Generalized Markup Language
TCP-IP	Transmission Control Protocol-Internet Protocol
USGS	United States Geological Survey
USNRC	United States Nuclear Regulatory Commission
WWSSN	World-Wide Standardized Seismographic Network

Appendix B
Minority Opinion

This report has a wealth of good material in it, but I feel that I must write a minority opinion on one main issue, the committee's recommendation to create the NSIR Federation. I think that the exact functions of the NSIR Federation are still not clear enough to immediately form it, especially since mechanisms to coordinate data activities already exist.

A group such as the NSIR Federation would not be a good method to set the hardware standards that are used in data systems (networks, tapes, etc.). The coordinated part of data directory efforts can be built around present interagency work. It is reasonable that NARA should request lists of datasets intended for long-term archival, but most of the process of evaluating datasets needs to be kept close to the working level. The discussion of standardization in the report should not be interpreted to mean that all agencies and archives should be forced to adopt certain standards and rework their data holdings into a common form and format. There are other concerns for which an analysis of the issues could be useful, but I believe that the NSIR Federation requires a better description of tasks and more debate before such a new body is established. Otherwise we may have more coordination, more systems, more cost, and less data.

Consider the important task of developing information about data. Information about datasets is needed in at least two or three levels of detail. At the highest level of information, the Master Directory methods that are in place for the GCDIS can be adopted (or even simplified more) to describe the datasets. This interagency Directory Interchange Format (DIF) is used nationally and internationally. We need to keep it simple enough so that people will submit the information. Some agency-level catalog efforts for datasets have existed since about 1968, and became more serious in the late 1970s. We should build on the GCDIS catalog efforts, and certainly not invent more complicated systems. Other data information efforts are needed, but they will be based on a bottom-up flow of ideas, on workshops, and the like. Each data system does not have to do exactly the same thing, but they must be easy to use. It is not clear that a formal NSIR Federation is needed to coordinate this.

How does the NSIR Federation relate to other data coordinating mechanisms? The Interagency Working Group on Data Management for Global Change (IWGDMGC) meets regularly to help coordinate data issues across many "global change" disciplines, which include air, water, ice, rocks, soils, and some biology. It seems to me that the IWGDMGC and the proposed NSIR Federation are mainly trying to do the same thing. They cover much of the same turf in terms of disciplines. They both want information about data, access to data, and data that will exist for more than 20 years. If we create separate organizations doing roughly the same thing, then it becomes even less likely that key agency people will attend the meetings. NARA asked the committee to consider how to deal with all the observational and laboratory physical sciences. The argument is made that we need an NSIR Federation because the IWGDMGC does not include some disciplines that were included in this study. However, NASA has control of most of the data for planetary sciences and astronomy so that this area may not be very hard to coordinate, except that data from ground-based telescopes should be included. This leaves the laboratory sciences, which can be handled as a special case.

Can the IWGDMGC be characterized as only agencies talking to agencies? No, there is a long-standing NRC panel, the Committee on Geophysical and Environmental Data, that has been asked to oversee its work, and that group has sponsored periodic national data forums. Does this mean that it is perfect? No, but it is not convincing to me that a roughly parallel coordination effort by an NSIR Federation would be necessary.

Some coordinating mechanisms besides the IWGDMGC will be needed to achieve the goal of making sure that long-term digital archives do exist, are adequately described, and can be used. NARA could hold periodic advisory panel meetings or workshops to talk about concerns and possible solutions. Similar issue-oriented meetings have been sponsored by other agencies and should continue. In interagency planning, the agencies should remember that some good data activities outside of agencies are funded by the agencies, but are probably not adequately represented by typical agency planning. This could be an argument for an NSIR Federation, but the problem could be handled in other ways. The idea of an NSIR Federation that is nimble, non-bureaucratic, and small is attractive, and it could even be a counter-weight to the agencies when that is needed. But we still have to ask: What would the NSIR Federation really do? Why would not it be just another coordinating office? Why would the agencies want to support it?

I believe that the NRC staff for this study has been very able and conscientious in helping to pull together this report. The report underwent significant change, but I was unable to fully support the committee's majority position regarding the proposed NSIR Federation. Nevertheless, I think that some divergence in viewpoints can help the sponsors and other readers to evaluate the best course of action.

Roy Jenne